T0064964

MASK of
the SUN

ALSO BY JOHN DVORAK

Earthquake Storms

The Last Volcano

How the Mountains Grew

MASK of the SUN

The Science, History, and Forgotten Lore of Eclipses

JOHN DVORAK

PEGASUS BOOKS
NEW YORK LONDON

MASK OF THE SUN

Pegasus Books, Ltd.
148 West 37th Street, 13th Floor
New York, NY 10018

Copyright © John Dvorak 2017

Firs Pegasus Books paperback edition April 2018
First Pegasus Books edition March 2017

Interior design by Maria Fernandez

All rights reserved. No part of this book may be reproduced in whole or in part
without written permission from the publisher, except by reviewers who may quote
brief excerpts in connection with a review in a newspaper, magazine, or electronic
publication; nor may any part of this book be reproduced, stored in a retrieval system,
or transmitted in any form or by any means electronic, mechanical, photocopying,
recording, or other, without written permission from the publisher.

Library of Congress Cataloging-in-Publication Data is available.

ISBN: 978-1-68177-668-2

10 9 8 7 6 5 4 3 2

Printed in the United States of America
Distributed by Simon & Schuster
www.pegasusbooks.com

To Sarah and Joyce,
who continue to enlighten my life

Contents

Note to the Reader

In designating years I have used the increasingly more common B.C.E. (before the Common Era) and C.E. (Common Era) instead of B.C. (before Christ) and A.D. (*anno Domini*). Also, to be consistent throughout the book, all dates are given according to the Gregorian calendar that is used today throughout most of the world, and not the Julian calendar that was in use at different times in the western world. For example, in 1715, when Edmond Halley was publicizing the total solar eclipse that would be seen from London that year, he gave the date of the eclipse as April 22 because England was still using the Julian calendar. The date I use is eleven days later, May 3, 1715, the date of the same eclipse according to the Gregorian calendar.

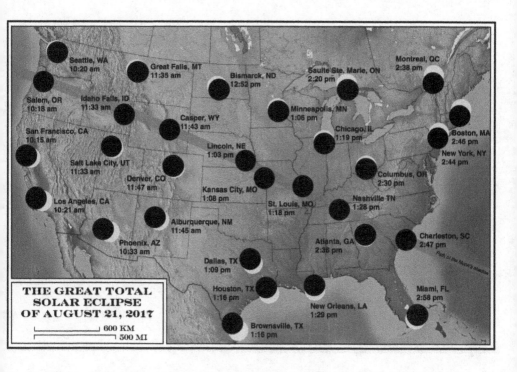

Path of the Moon's shadow across the United States on August 21, 2017. The maximum amount the Sun will be obscured by the Moon is shown for selected cities. The time of maximum obscuration for each city is also given.

Show me the eclipse, we say to the eye;
let us see that strange spectacle again.

—Virginia Woolf, after seeing a solar
eclipse from North Yorkshire,
England, June 29, 1927

PROLOGUE

New York, 1925

Eclipses be—predicted—
And Science bows them in—
But do one face us suddenly—
Jehovah's watch—is wrong.

—Emily Dickinson, 1862

I t was not clouds, but wind that was the concern.

For two days and nights, the giant dirigible *Los Angeles* had been kept inside a hangar to guard it from the violence of the wind outside, which blew crosswise, threatening to slam the *Los Angeles* against the doors if any attempt was made to emerge it.

The captain of this great airship, Navy Commander Jacob Klein, a veteran of the First World War, had skippered a destroyer that escorted troopships to France, routinely sailing through waters brimming with mines and menaced by German submarines. He had asked that a special weather forecaster be sent to the Naval Air Station at Lakehurst, New Jersey, where the *Los Angeles* was moored, to advise Klein when the weather might be improving, even slightly. As Klein knew, the *Los Angeles* would have to be launched before the next sunrise if he was to accomplish his latest assignment: to carry a dozen scientists aloft to observe a total solar eclipse.

Here it should be stated that the *Los Angeles* was a marvel of its age. It was the largest airship yet constructed. It consisted mainly of a giant airbag more than six hundred feet long and nearly a hundred feet high. A gondola hung on the underside with a command room in the bow and a large map room aft. In between were staterooms, rooms filled with bunks, two toilets, and a galley, so that up to four dozen people could be accommodated and comfortable onboard the *Los Angeles*. There was even a radio set so that those onboard the *Los Angeles* could keep in communication with those on the ground.

The giant airship had been built by the Zeppelin Company of Friedrichshafen, Germany, and given as part of the war reparations paid to the United States after the First World War as a condition of the Versailles Treaty. Its first official flight had been across the Atlantic Ocean in October 1924 when it was delivered to the Navy Air Station at Lakehurst. The second was a month later to Washington, D.C., where it received a christening from the President's wife, Grace Coolidge. The third, if weather permitted, would be to observe the total solar eclipse that would pass over the northeastern United States on January 24, 1925.

At 3:00 A.M. on the day of the eclipse, the weather forecaster woke Klein to tell him that there had been a noticeable drop in wind speed. The two men bundled themselves up and went outside. They stood there for almost an hour. The air temperature remained steady at a bone-chilling 5° Fahrenheit above zero.* But the wind speed did drop from twenty to twelve miles an hour. That was enough. Klein decided he would launch the *Los Angeles*.

The forty men who would fly with him that day, which included eleven scientists from the United States Naval Observatory in Washington, D.C., were awakened. Each man pulled on a heavy woolen shirt and woolen pants, an outer windproof jumpsuit designed for Arctic weather, a pair of wool-lined gloves, and a woolen cap with long flaps to protect the ears. When all were ready, the forty trudged outside, an onlooker thinking that they looked "like a fat man's regiment." Then, in turn, each man boarded the *Los Angeles*, climbing the short flight of metal stairs that led from the hangar floor up into the gondola.

Three hundred sailors who were stationed at Lakehurst were also awakened. They donned warm clothing, though not as protective as that worn by those who would fly that day. The three hundred then gathered inside the hangar where they were divided into two groups. A group of seventy-five was ordered to stand around the outside of the gondola, each man grabbing hold of a handrail. Their purpose would be to lift the entire airship and walk it outside. The remaining two hundred and twenty-five men were divided into subgroups of fifteen men each. Each subgroup was assigned a rope that hung down from the airbag. Their purpose was to keep the ropes taut and hold the *Los Angeles* close to the ground until given a signal to let it go.

As soon as everyone was in position, Klein gave the order to rev, in turn, each of the five gasoline motors that would drive propellers that would pull the *Los Angeles* through the air. After the last motor was checked out, the commander ordered the giant doors of the hangar opened. A gust of wind immediately swirled around the inside of the hangar. The men holding the ropes strained and steadied the airship. Klein then gave the order for the men holding onto the handrails to lift the *Los Angeles*, but the giant airship would not move.

---

* Minus 15° Celsius

The extremely cold weather had negated the calculations that were normally done to determine how much helium was needed inside the airbag to allow the *Los Angeles* to fly. Not having enough time to add more helium, Klein ordered ten men off the airship, including two of the scientists.

Now the airship could be lifted, and the seventy-five men holding on to the handrails proceeded to carry it outside. As soon as the *Los Angeles* cleared the doors, a gust of wind caught it broadside. There were struggles with the ropes, the men who were holding on to them skidding across a frozen ground. The seventy-five standing outside the gondola holding on to handrails were lifted several feet into the air. Klein had the motors revved and the giant tailfins trimmed. Slowly the giant airship settled back to the ground.

The two scientists got back onboard, as did two other men. That was enough. The dirigible was tugging skyward. The seventy-five let go of the handrails. The others released the ropes. The *Los Angeles* began to rise. At a thousand feet, Klein had the motors started and the airship began to circle the airfield. At two thousand feet, he ordered it turned to the northeast. The dirigible continued to rise. It passed over the communities of Lakewood and Asbury Park, the residents later recalling that they could hear the whine of the motors as the *Los Angeles* passed overhead. At six thousand feet it was over the coastline of New Jersey. Klein had the airship leveled off. They were now over open water, then over the coast of Long Island. Below was a frozen landscape. Ice outlined the many small bays and inlets. The Sun had risen and the sky was clear. The eclipse would begin in two hours. By then, making good headway, the *Los Angeles* would be at its destination over the eastern end of Long Island where those onboard would be ready for the eclipse—which would last almost two minutes.

Others were also preparing to view this unusual and spectacular cosmic event. It was later estimated that twenty million people, nearly one-sixth of the nation's population at that time, awoke early and positioned themselves to be somewhere along the path the Moon's shadow would follow, which extended in a narrow band from Minnesota to Rhode Island. To no one's surprise, the greatest number of people to see the eclipse assembled in the nation's largest city, New York.

And here one of the oddities of this particular eclipse is introduced. Though the path of the Moon's shadow would be more than a thousand miles long, its width when it passed near New York was barely seventy miles. And, as circumstances had it, the southern edge of the shadow would pass right through the city. Those who lived in lower Manhattan or in Brooklyn would not see the Moon completely obscure the Sun, while those who lived in upper Manhattan or Queens would. And so a mass migration occurred that day, one that began before sunrise, when hundreds of thousands of people trekked a mile or more north, riding in cars or taxis, in special trains or buses. Some went by foot. It was all so that they would be standing, even for a few seconds, within the complete shadow of the Moon. And so, by sunrise, every park and every public square in the northern half of the city was crowded with people. Every street corner that had an unobstructed view to the east was filled, people standing shoulder to shoulder. And every rooftop of every tall building had people clustered together and waiting.

In anticipation of how people's attention would be directed away from other matters, most stores and businesses announced in advance that they would open late that morning to give employees and customers a chance to view the eclipse. At Columbia University students were informed the previous day that all morning classes would be canceled. To guard against any criminals who might try to take advantage of the sudden darkness, Mayor John Hylan ordered all streetlights to remain lit until the eclipse was over. The warden at Sing Sing prison north of the city did something similar; he kept all prisoners locked in their cells that morning. And as a public service the city of New York arranged for a special radio show to be broadcast so that those who were blind or were unable to leave their homes could participate in the event.

The fact that the southern edge of the Moon's shadow would pass over the city's grid of streets provided an opportunity to do an experiment that had never been tried. Professor Ernest Brown of Yale University, the world's foremost authority on the Moon's motion, and hence of the timing and circumstances of eclipses, had done the calculations. He predicted that the shadow's edge would pass somewhere within a mile north or south of 110th Street in Manhattan. To test the prediction, Brown contacted Mayor Hylan, who talked with the head of the New York Edison Company, who agreed to have his company participate in the experiment.

One hundred forty-six employees of the Edison Company were divided into pairs. Each pair was assigned to stand on one of the rooftops of the long line of apartment buildings that ran along Riverside Drive on the west side of Manhattan. From those positions, each pair would have a clear view to the west of the Hudson River and the New Jersey shoreline. They would also have an unobstructed view of the Sun. One member of each pair was to watch for the approach of the Moon's shadow and decide whether it passed to the north or overhead. The other member was to gaze continuously at the Sun through a protective filter and decide whether the Moon ever completely covered the Sun or whether a thin crescent of brilliant sunlight was always visible.

On January 24, 1925, a Saturday, after eons of orbital motion, the Moon moved precisely between the Earth and the Sun, its shadow first touching the Earth just east of Upper Red Lake in Minnesota. Anyone who was standing there would have seen the Sun rise above the horizon as a black disk.

From there, the shadow raced eastward across the land, crossing over Wisconsin, Michigan, and southern Ontario. It passed over the northern half of Pennsylvania, across the central and southern portions of New York and, finally, over most of Connecticut and Rhode Island. From there, it ran out into the North Atlantic Ocean, the shadow leaving the Earth just south of the Faroe Islands.

The first community of any size to be darkened that day was Duluth, Minnesota, where the sky was overcast. One of the few people of Duluth who recorded the passing of the shadow was a farmer who, a few minutes earlier, had opened the door of his chicken coop letting out his chickens. According to the farmer, as darkness suddenly swept over them, the chickens at first seemed puzzled, then returned to the coop to roost.

The fact that this was the first total solar eclipse to pass over a broad region of the world that was connected by telegraph links provided an opportunity to conduct yet another experiment that had never been tried, that is, to measure the speed of the Moon's shadow. Telegraphic operators, chosen for their good judgment and quick reflexes, were stationed at Buffalo, Ithaca, and Poughkeepsie in New York and at New Haven in Connecticut, four cities close to the centerline of the shadow. On the morning of the eclipse, each operator was provided with a telegraphic key and an unobstructed view of the Sun. At the moment when the last ray of sunlight was hidden by the Moon, indicating the arrival of the shadow, that particular operator was to press the key sending an electrical impulse down a continuous string of wires to a special recording station located many miles away at the offices of Bell Telephone in Manhattan. Engineers who worked at Bell Telephone had devised a new mechanism that would record the arrival of each signal to within a tenth of a second. From those records, the actual speed of the Moon's shadow would be calculated.

Professor Brown had also done the calculations that predicted when each signal should arrive. The first signal, from Buffalo, according to Brown, should arrive at 6 minutes 24.5 seconds after nine o'clock. The actual arrival time at the offices of Bell Telephone was 6 minutes 23.7 seconds, eight-tenths of a second early. It was a remarkable confirmation of the theory of lunar motion then in use. At New Haven, the last station, the shadow arrived 5 minutes and 24.0 seconds after it had reached Buffalo. A map showed that the straight-line distance between the two cities was 320 miles. From that, the average speed of the Moon's shadow across New York and New England had been about 3,500 miles per hour, or nearly *a mile a second!*

During the last hours before the eclipse would happen, the mood in New York City was that of a grand party. Many people carried bottles of champagne. A few were still dressed in evening clothes, having spent the night in restaurants or clubs, planning to use the eclipse to end a night of revelry. At 7:10 A.M. the Sun rose. A cheer came up from the crowds scattered throughout the city. An hour later, those who had brought pieces of smoked glass or sheets of overexposed film could see that a small dark bite had been taken out from the side of the Sun, the first visual evidence that the Moon was beginning its slow progression in front of the Sun.

Not until nine o'clock had the sky noticeably darkened. By then, faces had taken on strange appearances. Dark yellowish-green circles appeared around eyes. Lips had turned dark purple in color. Only a thin crescent of sunlight was left. There were only minutes to go before the total solar eclipse would happen. Then, without warning, bands of light and dark started to shimmer over the crowds, looking, as one onlooker said, like a great ribbed shadow fence was passing over the ground. That lasted almost two minutes. Immediately after, some would remember feeling a brief gust of wind. Everyone sensed an increasing coldness. The sky darkened more. The Sun itself was now a sliver. Only hushed voices could be heard. Then, as the last ray of sunlight disappeared and darkness came over everyone, the silence was complete.

Many would later say it was the quietest moment in the city's history. No one spoke. Not a vehicle moved.

But the silence was short-lived. A pearl-white halo—the corona—with rays stretching outward now surrounded the Sun. And people started to talk again, at first in whispers, then in increasing volume until shouts could be heard. An applause rose, one that would rise and fall several times during the eclipse. Some people started to wave their arms around in a frantic manner. Others began to spin. Still others stood motionless and cried. Cars horns could be heard. There were thousands of them. People were leaning out of buildings and they were shouting and banging metal pans and anything else they could get hold of. And every church bell seemed to be ringing.

The celebration was citywide, and, yet, somehow, the small army of employees of the Edison Company who were standing atop the line of apartment buildings along Riverside Drive managed to do their assigned jobs.

When later interviewed, the member of each pair who was watching for the approaching shadow would report that no approaching shadow was seen. But the others, those who were to watch the Sun and see if it became completely extinguished, they had something certain to say.

Every person who had stood south of 96th Street reported that sunlight never completely disappeared; there was always at least a beam left. While those who were north of the same street reported seeing the Sun blanked out. The distance was later measured between the two observers who had been standing on opposite sides of 96th Street to be 225 feet. Though the Moon was nearly a quarter of a million miles away, the 225 feet was the maximum width of the edge of its shadow on the Earth.

After the eclipse, many people had an unexpected feeling of prolonged elation. Edward Riis, a correspondent for the *Brooklyn Daily Eagle*, who witnessed the eclipse, put it bluntly: "Nearly twelve hours after the event I am still struck with awe at the infinite majesty of the spectacle." He summed up the vision this way: "It was as if I had seen the hand of the Creator."

After the Moon's shadow passed over Manhattan, it was another second before it was over the eastern end of Long Island and the dirigible *Los Angeles*. The scientists onboard had made the last adjustments to their equipment. There was a battery of cameras loaded with a various types of film so that different aspects of the eclipse could be captured. There were electrical and magnetic receivers to record any sudden changes in those fields. The junior officers had been assigned to assist the scientists. They would record the readings of dials. Some had been given instructions on how to sketch the eclipse and were told to note whether any comets were near the Sun.

But the man who had the most challenging—and, in the opinion of those on onboard, the most enviable—task was Navy Chief Quartermaster Alvin Peterson. The Navy's best and most experienced aerial photographer, his assignment was to climb to the top of the dirigible and do what no one had ever done: take a motion picture of a total solar eclipse.

Access to the top was through a vertical shaft that ran the hundred feet from the top of the gondola to the top of the dirigible where there was a trapdoor that opened to the outside. Peterson made his way up by use of a ladder attached to the wall of the shaft. He carried a tripod, a movie camera, and several reels of film, including one that had been sensitized to low light that he would use during the darkness of the eclipse.

About an hour before the Moon's shadow would pass over him, he opened the trapdoor and set up his tripod and camera. The wind was a steady forty miles per hour. The air temperature was somewhere far below zero. Twice someone climbed up to check on him and offered to relieve him. But he wanted to see it through. In all, he would spend more than two hours standing on the top of the dirigible.

As soon as totality came, those in the gondola began their work. The scientists and their Navy assistants operated cameras. Someone called out the seconds to keep track of how much time had passed. The corona-sketchers began to sketch. Commander Klein, looking outside, would remember the scene succinctly as "a most spectacular sight." The sky overhead was a blue-black. All around, miles away at the horizon, beyond the limits of the shadow, was a flood of merging orange and red light.

Peterson was ready. He had switched to the reel of special film and was cranking the movie camera steadily. But it was at this moment, when the Moon completely blotted out the Sun, that the extreme cold finally gripped him. He braced himself against the cold and the sudden darkness. The loss of light meant he could no longer see the airship beneath his feet. Stars and planets could now be seen in the sky all around him, as well as a totally eclipsed Sun.

Later, back in the gondola, he would tell others "it was the weirdest sensation I have experienced." He had lost all sense of time and of place. He had been in an ethereal state. Though a strong wind had been blowing in his face, there were none of the common sensations of motion, no vibrations, no accelerations.

A doctor examined him and discovered he was severely frost-bitten across his cheeks and chin and on several fingers, something he had not yet noticed. His thoughts were still about what he had just seen. He was probably also thinking of the fact that, at the moment of the eclipse, when the great cosmic coincidence occurred and he was seeing the white light of the corona alone, he was in a privileged place: He was standing closer to the Sun than any of the millions who were watching from below.

It is simply a marvelous coincidence, nothing more, that, as seen from the Earth's surface, the apparent sizes of the Sun and the Moon are almost identical. If the Moon was 10,000 miles farther away or if it was a few miles smaller in diameter, it would not cover the Sun completely and there would be no total solar eclipses.

And total solar eclipses are not the only types of eclipses. The Moon's orbit around the Earth is elliptical. If it is near the far distant part of its orbit, its angular size is too small to cover the Sun completely, and so the tip of its shadow does not reach the surface of the Earth. In that case, there is an annular solar eclipse: The Moon passing directly in front of the Sun, but not obscuring it completely, leaving a ring of bright sunlight visible. And if, to an observer on Earth, the Moon does not cross directly in front of the Sun, but remains to the side, then it is a partial solar eclipse.

There are also eclipses when the Moon passes through the Earth's shadow. In those cases, if the entire disk of the Moon goes through the darkest part of the Earth's shadow, so that if one was standing anywhere on the Moon no direct sunlight would seen, there is a total lunar eclipse. Otherwise, it is a partial or a penumbral lunar eclipse.[*]

------

[*] This is detailed in the appendix: An Eclipse Primer.

If all of this makes it seem that eclipses are rare, they are not. If one lives for seventy years, one can expect to see about fifty lunar eclipses, nearly half of them total, and about thirty partial solar eclipses. If no effort is made to position oneself somewhere along the path of the Moon's shadow, then the chance of seeing a total solar eclipse wherever one happens to be living is about twenty percent.

It is the unusual appearance of the Sun or the Moon during an eclipse and the frequent occurrence of such events that leads to another set of statistics: There have been about 10,000 solar eclipses and about 6,000 lunar eclipses during the last 4,000 years, that is, during the period of written Western history—and this has made them invaluable to chronologists, geodynamicists, ethnographers, and solar physicists. It is the record of an ancient lunar eclipse that tells when the first Olympic games were played. And it is this record of an ancient solar eclipse that first showed that the rate the Earth is spinning is slowing by milliseconds per century. A mythical story about eclipses and incest supports the claim that the people of the Asian and American continents share a common heritage. And helium, the second most abundant chemical element in the universe, was first discovered to exist on the Sun thanks to a solar eclipse, and would not be found on Earth until decades later.

Yet despite the tremendous amounts of knowledge eclipses can yield, they are usually regarded as bad omens. They have commonly been associated with the deaths of kings or other royal personages, though such associations, when they have been made, are tenuous at best. It has also been suggested, in much more recent times, that eclipses precede drops in the stock market—and that may be real.

Eclipses have inspired poets and playwrights. Shakespeare mentioned both lunar and solar eclipses in *King Lear*. John Milton made an allusion to a solar eclipse in *Paradise Lost*, attributing the "disastrous" light of an eclipse to the tarnished image of the fallen Lucifer. Virginia Woolf traveled on an overnight train to see a total solar eclipse that lasted twenty-four seconds. William Campbell, director of Lick Observatory in California, once traveled to see a total solar eclipse that lasted one and a half seconds. Fans of the Boston Red Sox were ecstatic when, in 2004, their baseball team won the World Series for the first time in eighty-six years, some of the fans attributing the win to the mysterious force of a total lunar eclipse that hung in the sky during most of the game.

The lives of people as diverse as Alexander the Great, the slave Nat Turner, the British military officer T. E. Lawrence and Albert Einstein have been affected by eclipses. The British explorer James Cook timed the occurrence of eleven eclipses, both solar and lunar, during his three voyages around the Pacific Ocean. Those timings were used to establish longitudes of remote locations and, from that, produce better world maps. Mabel Loomis Todd of Amherst, Massachusetts, was one of the world's first eclipse chasers, making her travels during the late nineteenth and early twentieth centuries. And when she wasn't traveling to see an eclipse or lecturing at home about them she was doing another extraordinary thing—editing the poems of Emily Dickinson.

And that brings us back to Navy Quartermaster Peterson. His position atop the world's largest dirigible a mile above the ground gave him one of the most privileged views anyone has ever had of a total solar eclipse. But for him to be in the proper place at the correct time to see that particular eclipse required millennia of prior work to predict the exact motions of the Sun and the Moon. What follows is the story of how that long endeavor was accomplished, one that saw contributions from many cultures and that spans every historical age. Also told are many of the forgotten stories of how eclipses have influenced history—and how they are still important in our lives today.

MASK of the SUN

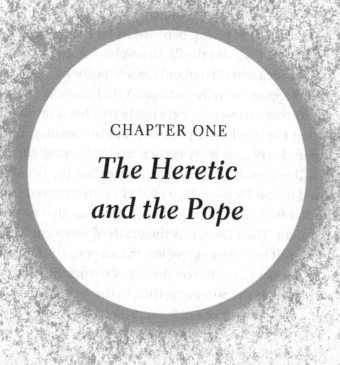

CHAPTER ONE

The Heretic and the Pope

Clouds and eclipses stain both moon and sun

—William Shakespeare,
Sonnet 35, c. 1600

B y the summer of 1626, during the third year of his papacy, Pope Urban VIII could be rightly pleased by all that he had accomplished. He had negotiated a peace treaty between the warring armies of France and Spain, then convinced the Catholic kings of both nations to turn their combined military might against the Protestant king of England, Charles I. He had ordered

Vatican troops to the nearby duchy of Urbino after the ruling duke died, thereby annexing that small state, making it easier for him to defend Rome. He had even managed to meet privately with his friend Galileo Galilei, giving permission for his fellow Florentine to publish a book that said the Earth revolved around the Sun, provided that the idea was offered only as a hypothesis. But there were matters that weighed heavily on Urban VIII. Foremost among them were rumors that astrologers were predicting his imminent death.

Predicting the death of a pope was not uncommon in Renaissance Rome. Forty-five years earlier, in 1581, amid rumors that Gregory XIII was in ill health, an astrologer had predicted that the pope would die on October 16, a prediction that proved false as he lived another four years, though in constant fear that he might die at any moment. Then there was the death of Innocent IX in 1591, the third pontiff in a row to serve less than a year. He died suddenly from an unknown cause. But on the day of his death, as everyone in Rome could see, as the Sun was setting in the west, a blood-colored full moon was rising in the east. Innocent IX had died just hours before of one of the most foreboding signs appeared in the heavens. He had died on the day of a total lunar eclipse.

The horoscopes that were produced during the Renaissance bear no resemblance to the highly stripped-down versions that appear in today's newspapers. Casting a horoscope was then a complex task. It required long detailed calculations. In fact, the job of an astrologer was synonymous with that of a mathematician. It was necessary to use every known branch of mathematics to determine, to great precision, exactly where the Sun, the Moon, and the five visible planets would be located among the stars at a specific time in the future. And people made important decisions based on such calculations. Urban VIII and his predecessors were among those who were firm believers in the usefulness and accuracy of astrology.

Urban VIII actually employed a small army of astrologers who produced horoscopes for the cardinals who lived in Rome, hoping that he would learn when one might soon be weakened by disease or might soon be struck by an unexpected death. When his own death was predicted, he took it seriously and had his own astrologers check the calculation. They confirmed the prediction, deciding that it was based on solid astrological reasoning.

The prediction was based on the occurrence of two eclipses that would happen in 1628. The first would be a lunar eclipse on January 20. The other and more auspicious one would be a solar eclipse on December 25 when the Sun and the Moon would be aligned in the constellation Sagittarius and the two warrior planets of Jupiter and Mars, which represented the pope, would be in close attendance.

Urban reacted quickly to the news. He ordered the removal of the time and the place of his birth—two elements that were crucial to casting a reliable horoscope of an individual—from all public records. He also knew that he could negate the ominous effects of the two eclipses if the proper countermeasures were taken. For that, he needed a master magician. And he knew where to find one.

Tommaso Campanella, a Dominican priest, had been in the prisons of the Inquisition for almost twenty-seven years. His troubles began in 1599 when, spurred by political unrest in his hometown, Stilo, located in the southern half of the Italian Peninsula—and knowing that the next year, 1600, was portentous by virtue of its numerical importance—it was a hundred times the addition of the two magical numbers of seven and nine—he led a rebellion against the Spanish rulers. But two of his conspirators betrayed him and he was captured and interrogated. Five times magistrates of the Inquisition authorized the use of torture to extract a confession from him. And five times he resisted. But because he had acted absurdly during his ordeal—for example, he had praised his inquisitors for their good work—and had spoken nonsensically for days after a torture session had ended, his inquisitors wondered if he might be insane. There was a special torture to test for that, the *veglia* torture, which was seldom used and deemed cruel even by the standards of the Inquisition. It involved tying the prisoner into unnatural positions, then hanging him from a ceiling and pricking him repeatedly with a host of pain-inducing instruments.

The rules of the Inquisition prescribed that the veglia torture be applied for no more than twelve hours, but the magistrates decided Campanella was a special case. And so he was subjected to it for thirty-six hours. At the end, his inquisitor, the two jailers, and the two doctors who were present during the torture filed reports. The reports were read and discussed by the magistrates who concluded that, because Campanella had endured the treatment without confessing or without calling for mercy—signs that all trace of God had left him—he must be insane. Now, also according to the rules that governed the Inquisition, his insanity established, Campanella could never again be tortured nor could he be executed. Instead, he would spend the remainder of his life living in the dungeons of the Inquisition.

During his years of confinement, Campanella studied astrology, learning how to cast horoscopes, which he did for other prisoners and for some of his jailers. Once in 1618 his jailers led him outside, hoping the vision would improve his ability to make predictions. He also wrote several books. In one he argued that some forms of magic, which he claimed to know, could counter the negative effects produced when heavenly bodies were aligned in unfavorable positions. That of course brought him to the attention of Urban VIII who ordered Campanella brought to Rome.

He arrived in the city in secret, dressed as a peasant. He was kept in a prison cell at the Vatican and was taken, several times, to the papal residence at the Quirinal Palace near the Vatican to meet with Urban and to perform his magic. During the meetings, the two men were dressed in white robes. They sealed themselves inside a dark room to be protected from the evil contents of outside air. To cleanse the air within the room, Campanella sprinkled rose vinegar and burned incense made of laurel, myrtle, and rosemary. "Nothing has more effect against the power of the stars," Campanella would write, "even if the poison has been administered diabolically." He then hung cloths of white silk on the walls, symbols of purity. He lit two lamps and set up five torches that represented the Sun and the Moon and the five planets. On the walls, Campanella drew the twelve signs of the zodiac. He took out gemstones associated with the beneficent planets Jupiter and Venus. He and Urban drank liquor that had been distilled under the influence of the two planets. Then the two men knelt and Campanella led long prayers while the two men listened to soft music played by musicians who were outside the room. The music was the most direct way to disperse the pernicious qualities of eclipse-infected air.

The ministrations proved successful. The lunar eclipse on January 20, 1628, was visible from all of Europe, and nothing harmful happened to Urban. From Rome, the solar eclipse in December occurred when the Sun was low in the western sky, lessening its influence. Moreover, the silhouette of the Moon never passed directly in front of the Sun; instead, only a small part of the Sun was ever obscured. And so, again, Urban survived. On January 11, 1629, in recognition of his service to the Holy See, Urban released Campanella from prison. A month later, the head of the Dominican order in Rome bestowed on him the title *Magister Theologiae*.

But soon another prediction of the pope's imminent death was made. And it too involved an eclipse. The prediction was made by Orazio Morandi, abbot of the monastery of Santa Prassede near Rome and the most respected astrologer in the city.

His prediction was based on yet another solar eclipse, one that would occur in little more than a year on June 10, 1630. Sensing an opportunity, the Spanish ambassador to Rome, Cardinal Gaspar de Borgia, sent a message to Madrid requesting all cardinals in Spain to come to Rome and prepare for a conclave to elect Urban's successor. The French and German ambassadors did likewise, sending similar messages to their countries, fearing that if they delayed, they and their cardinals would play no role in choosing a new pope.

Urban now felt besieged, so much so, that he moved to the papal retreat at Castel Gandolfo several miles from Rome, where security was so tight that even servants and courtiers had trouble getting past the guards. Urban sent for Campanella once more and asked him to perform his magic.

Again, the two men met in a sealed room. Rose vinegar was sprinkled and incense burned. Silk cloths were hung. Two lamps and five torches were lit and signs of the zodiac marked on the walls. Then Urban and Campanella said long prayers.

And, again, the counter-magic worked. As we know today, from being able to calculate the Moon's past motion accurately, the track of the solar eclipse that day began in Canada, traveled across the Atlantic, and then over France. The obscured Sun was last viewed at sunset on the island of Corsica. An eclipsed Sun, even a partial one, was never seen from Rome.

On July 15, five weeks after the eclipse, Morandi was summoned to appear before the magistrates of the Inquisition. His interrogation began a week later. In November the former abbot was found dead in his prison cell. The physician who examined the body was careful to report that there was no evidence of foul play; instead, he concluded that Morandi had died of a fever. Given the political intrigue of the era, few people then, or now, believed the physician's conclusion.

Soon after Morandi's death, Urban took action in another way, one that would be a landmark in the history of the Catholic Church. To understand the impact, one must remember that since ancient times few doubted that the heavens influenced activities on Earth. By the mid–thirteenth century, astrology had been integrated as part of the curriculum of Western universities, as part of the quadrivium alongside arithmetic, music, and geometry. And it had been firmly allied with medicine since the fifteenth century, with medical chairs in astrological divination existing at universities all across Europe, including at Bologna, Naples, and Paris. Several popes had been patrons of astrologers. But the attitude was changing. The Protestant reformer Martin Luther criticized astrology as a dangerous game with the devil. Galileo and others were showing how the natural world could be understood independent of astrological methods.

In his own way, Urban was showing himself to be a progressive. In 1624 he made the use of tobacco in churches punishable by excommunication. In 1628 he forbade the enslavement of native people at Catholic missions in South America. He also had something to say about the future use of astrology.

On April 1, 1631, he issued the papal bull *Inscrutabilis judiciorum Dei*, "The Inscrutable judgments of God," that prohibited any member or any official of the Catholic Church from engaging in astrological predictions. It was a sweeping condemnation of the art that, in Urban's words, dared "in its sinful curiosity to pry into the mysteries which are hidden in God's heart." No longer could someone within the Church predict the death of a pope from astral influences. And, indeed, Pope Urban VIII lived a long time, until 1644, his term as the Holy Father among the longest, lasting twenty-one years.

Campanella never considered the bull to apply to him. He continued to cast horoscopes as a way, so he wrote, to divine the "angelic intelligences" that God had placed among the stars. His persistence got him into trouble in Rome with clerics who could cast a blind eye for only so long. And so he left the city and went to Paris where he and his craft would be tolerated, finding a place at a Dominican convent. But the influences of the Sun and the Moon were not kind to him. In late 1638 he realized that a solar eclipse that would occur on June 1 of the following year would be fatal to him. He reenacted, in the privacy of his room at the convent, the ritual he had performed for Urban. This time the counter-magic did not work. Campanella died in his bed eleven days before the eclipse.

This short episode in the history of the Catholic Church continues to have its influence today. In 1998, concerned about the public's increasing interest in the occult, Pope John Paul II sent a letter to all bishops in which he reminded them of the existence of *Inscrutabilis* and that "esoteric superstition found in astrological speculations" was "incompatible with the Christian faith."

Here it is intriguing to wonder what Urban VIII and Campanella, as well as the Spanish astrologer Morandi, would have thought of the fact that a future pope, John Paul II, one of the most revered pontiffs in Church history, was born and was buried on days when there were solar eclipses.

_____

But it is not just popes or Catholics that have paid special heed to eclipses throughout history. Eclipses have been almost universally viewed as harbingers of misfortune or catastrophe. One needs to look no further than Shakespeare for examples.

"These late eclipses in the sun and the moon portend no good to us," declares the Earl of Gloucester in *King Lear*. Antony complains in *Antony and Cleopatra*: "Our terrene moon is now eclipsed." That alone, he says, foretells his own fall. Othello decides that "a huge eclipse of sun and moon" would be a fitting accompaniment to the terrible tragedy of Desdemona's death. And, in *Hamlet*, Horatio, speaking of the several evil omens that foretold Caesar's death, refers to the Moon as "sick almost to doomsday with eclipse."

Less than fifty years after Shakespeare's death in 1616, the most popular publisher of almanacs in England, John Gadbury, keeping with the tone that eclipses were bad omens, made repeated references to a solar eclipse that would be seen from Ireland and Scotland. The event occurred on a Monday, April 8, 1652, a day that became known, forevermore, in British history as "Black Monday" because of the suddenly darkening of the sky. The event also created a great deal of anxiety among people who lived along the eclipse path, many of whom saw it as the beginning of God's wrath to the Day of Judgment, the refrain of a ballad composed in anticipation of the eclipse recommending: *"Repent therefore O England/The day it drouth near."* Gadbury would write that the eclipse "poured down its influences so violently and cruelly." He also reminded readers of his almanacs that solar eclipses presaged "the death of Kings and Great persons, alterations of Governments, changes of Laws." If, at the time Gadbury was publishing his almanacs, one examined recent events in English history, one would discover that it did indeed seem to be true.

There was Queen Anne Neville, one of the least conspicuous monarchs in English history. There are no reliable portraits of her, and none of her letters survive, assuming that she wrote any. In fact, chroniclers living at that time seldom mentioned her. She died on March 16, 1485. On that day, the path of a total solar eclipse passed over France and central Europe. Even that fact would probably have gone unrecorded except for one important thing. The death of her husband, King Richard III of the House of York, a few months later also involved an eclipse.

Richard III died at the Battle of Bosworth, the last battle of the Wars of the Roses. The victor was the man who would become King Henry VII, the first of the Tudor kings. He had the heavily wounded and naked body of Richard III put on display under the arches of the Church of the Annunciation in Leicester. A full moon occurred on the third and final night the body would be displayed. The body was placed so that it would be illuminated by the light of the full moon, but, as the night progressed, a lunar eclipse occurred and the moonlight was noticeably dimmed, a confirming signal, so many thought, that the influence of the House of York had passed.

Another notable association of a royal death and an eclipse is recorded in the *Anglo-Saxon Chronicle*, compiled originally on the orders of King Alfred the Great in the ninth century and maintained for nearly two hundred years by generations of mainly anonymous scribes. The chronicle of interest here was written in the twelfth century by John of Worcester, a monk of the priory at Worcester Cathedral in England. For a scribe, he did not have the most disciplined or orderly hand. For instance, he would indicate in margins which text was to be erased or substituted, then not bother to erase his notations. Nevertheless, his chronicle is now valued most for its contemporary account of the reign of Henry I.

According to the chronicle, on August 2, 1133, a Wednesday, during the thirty-third year of his reign, Henry, the English king, a son of William the Conqueror, surrounded by his usual retinue of knights, was preparing to leave the English coast and sail for Normandy. Suddenly what seemed to be a dark cloud appeared overhead. The king and his followers walked about, marveling at it. They raised their eyes skyward and saw the Sun shining as if it was a new moon. But it did not keep the appearance for long. The curve of sunlight narrowed until all the light went out, making it necessary to use candlelight to do anything. Stars also appeared. Then the sequence in the sky reversed as a sliver of sunlight reappeared, growing in width until the full Sun was again visible. Many said a great event would come. King Henry crossed the channel. And a great event did happen: The king died.

His death did not come immediately. He died two years later in France from eating lampreys. But what was important to his people was that he never returned to England. And many attributed that to the ominous occurrence of a total solar eclipse on the day of his departure.

Though such broad associations seem quaint today, they were commonly made throughout history and in different cultures. The Roman historian Tacitus noted that the death of Emperor Augustus on August in the year 14 c.e. occurred one month before a lunar eclipse. An annular solar eclipse across France on May 5, 840, presaged the death of the ruling king, Louis the Pious, son of Charlemagne, forty-six days later. King Olaf II was killed at the Battle of Stiklestad on July 29, 1030, one of the most famous battles in the history of Norway. Thirty-three days later, according to the *Sagas of the Norse Kings*, though the weather was clear and the Sun shone brightly, the day suddenly "became dark as at night." The darkening was a solar eclipse, and, though it came more than a month later, it was still associated with the death of a king.

And such associations are not limited to Western cultures. According to the *Nihongi*, the oldest history book in Japan, Empress Suiko, who had reigned for thirty years and had brought Buddhism to Japan (making her one of the most important figures in Japanese history), died five days after a solar eclipse in the year 628. King Juuko of Buganda, modern-day Uganda, died, according to oral tradition, when "the sun fell out of the sky," an event that many scholars interpret as a solar eclipse in 1680. On North Island, New

Zealand, a lunar eclipse was seen at moonrise on May 1, 1836, and was considered by some people as a sign that a leader would die. In fact, two did, Kiharoa and Hikareia of the Mataatua tribe.

And so it should come as no surprise that, before its occurrence, when the path of a total solar eclipse was predicted to pass over London in 1715—the first such eclipse to do so in nearly 600 years—there was much consternation. Months in advance, in broadsheets devoted to the subject, the day of the eclipse, May 3, 1715, was being proclaimed "The Black Day or a Prospect of Doomsday." As people were thinking: Were not both the execution of Charles I in 1649 and the Great Fire of London in 1666 preceded by a few months by eclipses? (Both were.) Just 230 years earlier, the deaths of both Queen Anne and King Richard III had been associated with eclipses. And what of the current reigning monarch, King George I, of the House of Hanover, who had ascended to the throne just months before the predicted eclipse? Should people be concerned about his safety? More to the point, should the coming eclipse be viewed as a celestial judgment of the newly crowned king?

The eclipse came and went, and George I survived unscathed, though where he was and whether he saw the spectacle remained unrecorded. He did continue to live and rule for another twelve years. But his counterpart in France was not so fortunate.

Louis XIV had been known as *Le Roi Soleil*, "The Sun King," for much of his reign. It originated when he was a young man and had appeared as the sun god Apollo in *Le Ballet de la Nuit*, "The Ballet of the Night." And so it seems fitting "The Sun King" was extinguished soon after a total solar eclipse.

He did see the eclipse, which was a partial solar eclipse throughout France, though total in London. Several weeks later, he complained of a pain in one leg. The pain developed into gangrene. On September 1, 1715, after weeks of agony, the man who had chosen the Sun as his emblem and whose name is synonymous with absolute power, died inside the grand palace he had built at Versailles.

The death of a monarch was not the only possible evil consequence of an eclipse. For most people, the greater concern was
the spread of disease.

The first well-documented outbreak of syphilis was in Europe in
1495. It erupted among mercenaries and prostitutes associated with
the French army of Charles III that was trying to conqueror the port
city of Naples. For that reason, it was soon known as the "French
disease," though it must be noted that two years earlier Spanish
physician Rodrigo Ruiz Díaz working in Barcelona reported that he
treated sailors who had the horrible genital skin eruptions that is
characteristic of the disease.

As to the cause, many well-educated people knew the reason. Joseph
Grünpeck of the royal court of Austria, who contracted the disease in
1501 and devised an early treatment, saw the origin in the near coincidence of two celestial events. The first was a Grand Conjunction,
that is, a close meeting of Jupiter and Saturn in the sky. It occurred on
November 25, 1484, at a time when both planets were in the part of
the sky represented by the sign of Scorpio. As Grünpeck and his contemporaries reasoned, because Scorpio ruled the genitals—each sign
of the zodiac corresponds to a part of the body—the first lesions of the
disease appeared in that area. The Grand Conjunction had provided the
poison. Its release came a few months later on March 16, 1485, when
a total solar eclipse was seen from Europe.* But why had it taken ten
years for the poison to reach Earth? Grünpeck and others had reasoned
that it had taken that long to travel the great distance from Jupiter.**

---

* This was the same eclipse that occurred on the day of the death of Queen
Anne Neville of England.

** Today our view of the origin of syphilis in Europe is that it was almost certainly contracted by men who sailed with Christopher Columbus in 1492 on his
first voyage to the New World, which accounts for its appearance in Barcelona
the following year. Furthermore, some of the mercenaries in the French army
who laid siege to Naples in 1495 may have sailed with Columbus. As a final
note, molecular geneticists who have sequenced the DNA of different strains
of the bacterium that causes syphilis conclude that the strains found in Europe
probably originated in the Caribbean.

As another example of the association of an eclipse and disease, in 1585, Englishman Thomas Harriot observed a partial solar eclipse while crossing the Atlantic en route to the colony on Roanoke Island off Virginia. Soon after his arrival, he noticed that many natives died of a mysterious disease after coming into contact with him or other Englishmen. He attributed the tragedy ultimately to divine intervention, though he acknowledged in his writings that the deaths had been foretold by "a terrible solar eclipse" that had occurred on the voyage over.

James Lind of the British Navy, known today for being one of the first to demonstrate the effectiveness of citrus fruit against scurvy, was in the Bay of Bengal in 1762 when he witnessed the deaths of hundreds of Europeans and tens of thousands of local people. According to Lind, the greatest number of deaths occurred on the day of a solar eclipse, October 17. "This fever was so general on the day of the eclipse," Lind wrote, "that there was not the least reason to doubt its effect."

As a final example, on June 8, 1918, a total eclipse was seen across the United States, the path of totality running from coast to coast. A worldwide epidemic of influenza began in the fall of the same year. It was the most devastating disease in history, killing more people in a single year than the Black Death had killed in a century during the Middle Ages. About one-fourth of the 1.8 billion people then living were infected, and more than fifty million died from the disease or its complications. Naturally, some populations suffered more than others. In the United States, Native Americans living in the Southwest were struck particularly hard.

In stark numbers, of the 30,000 Navajos still living in the United States in 1918, nearly 4,000 died, which is a mortality rate of about twelve percent. As one survivor remembered: "The sickness hit us in the fall; it started spreading across the reservation almost overnight and lots and lots of people died from it. People would feel fine during the day, get sick in the night, and by morning, they'd have all passed away." But what was the reason for this catastrophe? The survivor continued: "Even the people trained to find out the causes of illnesses . . . couldn't determine what was happening."

In Navajo culture, as in many cultures, people do not think tragedies simply happen. Signs appear beforehand, though they may not be recognized until after the fact. And so it was with the 1918 epidemic and the devastating impact it had on the Navajo population. To some, after reflection, the catastrophe was presaged a few months earlier by a solar eclipse that had been visible across the American southwest on June 8, 1918.

To those who pride themselves on rational thought, the fact that the deaths of some monarchs can be associated with some eclipses is not a surprise: On any given day, including one when an eclipse happens, some people died. And so it would be surprising if there were *no* examples of monarchs dying on or close to days of eclipses. Likewise, today we have a much better idea than we did a hundred years ago as to how diseases originate and how they spread. And so, again, it would be surprising if *no* epidemics could be associated with eclipses. We think of such associations as relics of a less enlightened past, of a time when we were less knowledgeable and more prone to superstition. But, if we turn the focus on ourselves, we find that much of our behavior is still irrational—and that we still have ingrained within us the idea that eclipses are bad omens. As a case in point, consider the negative impact that eclipses have on the stock market to this day.

—————

Gabriele Lepori is a financial behaviorist, that is, someone who uses psychology to explain behavior in the stock market. In 2009 he wanted to test whether the perception of a bad omen might influence whether stock traders decided to buy or sell stocks. He chose eclipses as the bad omen because they occur suddenly and predictably. To his great surprise, he concluded there was such an effect.

But, first, let's step back. As Lepori pointed out in his study, it is no secret that certain groups of people, such as athletes or gamblers, tend to engage in superstitious activities. They may wear old hats or socks or eat specific foods or wear special amulets, believing these will increase their chance of success. Many stock traders do the same. Furthermore, as many anthropologists have pointed out, such as Bronislaw Malinowski, one of the most highly acclaimed and influential anthropologists of the twentieth century, superstitions are actually necessary for human existence. They are used to fight anxiety and distress by filling the psychological gap caused by uncertainty. And so those who work in environments that are highly competitive and stressful and where outcomes are uncertain, these people tend to be the most superstitious. And that includes stock traders.

In his study, Lepori considered all 362 eclipses that occurred during the eighty years between 1928 and 2008. He examined stock indices and, after considerable statistical analysis, concluded that stock prices generally fall on days of eclipses. He also noticed that the volume of trading was usually low on those days, indicating that traders were holding back. Both the fall of stock prices and the low volume suggested that traders were cautious to do business on such days. Furthermore, the drop in the stock indices was slight, usually less than one percent, and the value of the indices recovered within three days.

I decided to spot-check Lepori's conclusion by examining a highly selective set of eclipses: Those that I had witnessed and had been memorable. The first was a solar eclipse. It occurred during the winter, and I was lucky when a small bank of clouds moved away from the Sun halfway through the eclipse so that I could see the corona. On that day, February 26, 1979, the Dow Jones Industrial Average dropped twenty points. Two days later, the Dow recovered and rose twenty points.

I witnessed a solar eclipse from the Hawaiian Islands in 1991. On that day, the Dow Jones Industrial Average rose fifteen points and continued to rise over the next several days. A third solar eclipse that I saw, this one seen from north Texas, coincided with a drop of thirty points, followed two days later by a full recovery.

As for lunar eclipses, the most recent one that is memorable in my mind occurred on August 28, 2007, when a bright full moon turned a deep copper color. On the day before the lunar eclipse, the Dow dropped three hundred points. It dropped another two hundred points the day of the eclipse. That was a Friday. The next business day, a Monday, the Dow rose two hundred points.

As already pointed out, this is a highly selective list of eclipses, and yet it lends some credence to Lepori's conclusion. And so I decided to test a completely different type of celestial event.

In mid-August 2010 there was a close meeting of the planets Mars, Venus, and Saturn with a new moon in the western sky. The Perseid meteor shower also occurs in mid-August. On the night it peaked that year, August 12, I saw more than eighty meteors in an hour. I have since checked the stock market and saw that on the day before the meteor shower the Dow dropped three hundred points and continued to drop during the next three days, the duration of the shower. It is an intriguing result, but there is something else hidden here.

As Lepori has pointed out, not every eclipse—and I will add not every meteor shower, which many cultures also consider to be a bad omen—is associated with a drop in stock prices. Instead, it is those that are highly publicized beforehand that seem to have an effect. Stock traders are reacting, either consciously or subconsciously, to what is generally regarded as a bad omen, producing a self-fulfilling prophecy.

———·———

Whether one ascribes psychological forces—or, even, astrological ones—as the cause for a drop in the stock market when an eclipse occurs or whether one challenges that any relationship between the two exists, what is fascinating is the sheer longevity and scope this ingrained fear of eclipses has had throughout world history.

The earliest written record of an eclipse that actually gives the date when the eclipse occurred appears on a clay tablet found in

1848 among the ruins of the ancient city of Ugarit in what is now Syria. The inscription begins: "The day of the new moon in the month of Hiyar was put to shame. The Sun went down (in the daytime) with Mars in attendance."

Because one knows the time period covered by Ugaritic texts, from about 1450 to 1200 B.C.E., and if one accepts that the inscription refers to a solar eclipse that occurred "in the month of Hiyar," then one can determine that the inscription on the clay tablet must be recording a solar eclipse that occurred on May 3, 1375 B.C.E. Even more intriguing is the statement at the end of the inscription.

It is a warning: "The overlord will be attacked by his vassals." And so even at the beginning of history eclipses were regarded as bad omens.

But there is a prudent way of dealing with a bad omen. As Pope Urban VIII understood, if one knew when an eclipse was about to occur, then the appropriate counter-magic could be performed.

And that required having a reliable way to predict them.

CHAPTER TWO

The Invisible Planets
of Rahu and Ketu

The Sun was completely engulfed by Rahu.

—*Ramayana*, an epic poem of India,
c. seventh century C.E.

The history of the region around the city of Bhubaneswar in India, located on the edge of a wide delta that extends into the Bay of Bengal, can be traced for more than two thousand years. During those millennia hundreds of Hindu temples were built. A few are huge, covering several acres, but most are modest and consist of one or two architecturally impressive structures. One of these, known for the façades of intricate stone carvings that cover every square inch of the inner and outer walls, is the temple of Mukteshvara, that is, the temple "to the Lord who provides freedom of mind through the practice of yoga."

To approach Mukteshvara one first has to pass beneath an arched gateway held up by two thick pillars, each one decorated with carvings of smiling goddesses with ample breasts, their bodies adorned with numerous strings of beads and other ornaments. Once through the gateway, one then walks across a small courtyard and, after lowering oneself, enters the first of two buildings, the outer sanctum.

Inside the wall carvings are mostly floral designs interrupted by repeated patterns of swirled lines. The figures of holy men look down from the corners of the ceiling. After passing through and exiting this room, one crosses a narrow open space and enters a second building, the inner sanctum. This is the most sacred part of the temple.

The walls of the inner sanctum are also richly adorned with floral designs, swirled lines, and stylized figures of holy men and voluptuous women. But in this room there is something distinctly different. If one turns after entering and looks over the doorway, one sees a carefully carved stone block. And carved into the block is a row of nine human figures, each one within its own shallow niche. These nine figures are the *Navagrahas*, "the nine planets."

As examined from the left, the first seven are almost identical in form: a human figure seated with legs crossed in a lotus position. The leftmost one represents the Sun. The next is the Moon. The remaining five are the five visible planets that, though star-like, move across the sky relative to the fixed stars. In order, from left to right, they represent Mars, Mercury, Jupiter, Venus and Saturn. But what of the other two figures? One is simply a giant human head; the other is an upper human torso with hands pressed together and fingertips pointed upward. These, too, are important objects that move across the sky relative to the stars. They are the invisible planets of Rahu and Ketu. They are the celestial demons that cause eclipses.

To understand the origin of Rahu and Ketu, one must understand one of the origin stories of Hindu mythology, the story of the "Churning of the Ocean."

Long ago, when only gods and demons existed, both were desirous of obtaining the elixir of immortality that was contained within a pot that lay on the bottom of the sea. Because neither the gods nor the demons had sufficient power to retrieve it themselves, they made a pact. They would work together to churn up the ocean and, in that way, retrieve the pot.

With their dual forces, they uprooted a mountain and used it as a churning rod. They rotated the mountain by using a giant snake as a rope. After much work, the ocean was churned and the pot with magic elixir rose to the surface. The pot was taken up by Vishnu, one of the supreme gods, who was then in the form of the beautiful maiden Mohini, literally, "the one who enchants." She instructs the gods and the demons to sit in two parallel lines so that she can distribute the elixir. She begins by giving it only to the gods. Most of the demons do not notice this because they are charmed by Mohini's beauty, but the great demon Swarbhanu realizes the trick and, changing lines, quietly goes and sits with the gods.

Mohini has given him a few drops of elixir when the gods representing the Sun and the Moon recognize Swarbhanu and warn Mohini that she is feeding a demon. She becomes enraged, and with a sword decapitates Swarbhanu, his head becoming the new demon Rahu and his torso Ketu. For vengeance, Rahu and Ketu chase the Sun and the Moon across the sky, occasionally gulping them down, but, because they are only a head or a torso, the ingestion is not permanent. And so eclipses can only be temporary.

There are details of the Rahu-Ketu story that show that the originators of the story had a sophisticated understanding of the movement of the Sun and the Moon across the sky. The originators of course understood that the Sun and the Moon move across the star-filled heavens from west to east.* They also portrayed Rahu and Ketu as cunning because the two demons moved across the sky in the opposite direction, from east to west, making it easier to capture and gulp down the two luminous orbs that betrayed Swarbhanu and denied the demons the elixir of immortality.

But to understand what this means astronomically it is necessary to understand the basic geometry of eclipses.

The annual path the Sun follows across the sky is called the "ecliptic," derived from a Latin word that means "of an eclipse" because it is only along the ecliptic that the Sun and the Moon can meet and an eclipse can happen. The Moon follows a path that is always close to the ecliptic, but does not exactly correspond to it; though the Moon does cross the ecliptic twice each time it orbits the Earth, that is, about twice every month.

Now let's dig deeper.

Imagine you are standing at the center of a hollow sphere. On the wall of the sphere are all of the stars and the Sun and the Moon. The Sun and the Moon follow paths that are two great circles that are nearly coincident, but not quite. And that is important. If the two paths did coincide, then there would be a solar eclipse and a lunar eclipse every month; that is, there would be a time every month when the Moon passed directly in front of the Sun and when the Moon was directly opposite the Sun. But the two paths are inclined slightly by 5°, which means they cross at two points. And it is only when the Sun and the Moon are *both* near one of these two points, known as nodes, that an eclipse can

---

* The Sun and the Moon rising in the east and setting in the west is an illusion caused by the Earth's rapid rotation. To satisfy oneself that the Moon actually does revolve around the Earth from west to east, look at the Moon one night and note its position with respect to several nearby bright stars. Then, an hour later, look at the Moon again. Its position will have shifted *eastward* of those same stars.

happen: If they are both close to the same node, then there is a solar eclipse; if they are at opposite nodes, then a lunar eclipse happens.

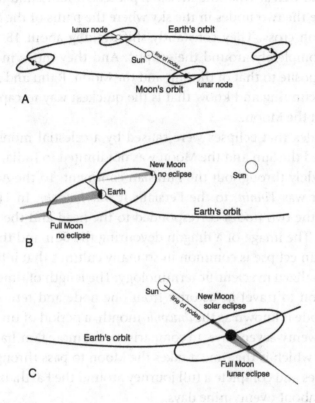

(A) Because the Moon's orbit is inclined to the Earth's orbit, eclipses can occur only when the Earth is near one of the lunar nodes. (B) If the Earth is not near a lunar node, then no eclipses can occur because, at the moment of new moon or full moon, the Moon is too far from the plane of the Earth's orbit. (C) If the Earth is near a lunar node, then eclipses occur because the Sun, the Earth and the Moon all lie along a line during new moon and full moon.

Furthermore, because the motion of the Moon is influenced not only by the gravitational attraction of the Earth, but also by the attraction of the Sun, the Moon's path across the sky shifts slightly—and the positions of the two nodes shift accordingly. The

Sun and the Moon both move across the sky from west to east (as already mentioned), and the two nodes shift in the opposite direction, traveling from east to west.

It is now clear what the invisible planets of Rahu and Ketu are: They are the two nodes in the sky where the paths of the Sun and the Moon cross. They constantly shift, taking about 18 years to move completely around the ecliptic. And they move in a direction opposite to that of the Sun and the Moon. Rahu and Ketu are indeed cunning and know that is the quickest way to capture the Sun and the Moon.

The idea that eclipses were caused by a celestial monster that devoured the Sun and the Moon was not limited to India, but was held widely throughout the Eurasian continent. To the Arabs the monster was *Tinnin*; to the Persians it was *Jawzahr*. In Tibet and China, the two nodes corresponded to the head and the tail of a dragon. The image of a dragon devouring the Sun and the Moon during an eclipse is common in so many cultures that it has been memorialized in scientific terminology: The length of time it takes the Moon to travel in its orbit from one node and return to the same node is known as the *draconic* month, a period of time that is about twenty-seven days. In comparison, the more familiar *synodic* month, which is the time it takes the Moon to pass through all of its phases and complete a full journey around the Earth, is slightly longer, about twenty-nine days.

———•———

The ability to track the movement of the two lunar nodes is just one of the elements that can lead to eclipse prediction. Another is to know when the Sun and the Moon would be close to a node. But how easy is it to follow these two luminous objects? For the Sun, it is very easy.

At some time far off in prehistory, people realized that the passing of the seasons was related to where the Sun rose on the horizon. This was of importance because, by following the Sun's movements, they would know when to migrate to find animals to hunt or when to expect rivers to fill with fish or when to plant crops. In particular, three points on the horizon were noted: the two extremes that marked the farthest north or south the Sun would rise and the point between them. Today we know the extremes as the *solstices*, the days that mark the beginning of summer and winter, and the midway point as the *equinox*, which is the beginning of spring or fall.

The equinox is especially easy to determine. One needs simply to erect a pole or place upright a slender rock, then watch where the long shadow points at sunrise. If it points directly at the place on the horizon where the Sun will later set, that day is the equinox. Furthermore, the line defined by the long shadow at sunrise on the day of the equinox is precisely an east-west line. And so no extraordinary knowledge of the Sun's movement or the ability to conduct sophisticated surveys is required to give an east-west alignment. Which may explain why such alignments are not unusual. Among the most famous are the pyramids of Egypt. While the actual construction of these great edifices is praiseworthy—and still not totally understood—the alignment of their bases to the cardinal directions did not require superhuman effort—or the advice of alien astronomers, as some have maintained—merely the persistent observation of the rising of the Sun.

To establish a line to a solstice is only slightly more complicated. One can use the same pole or upright rock and note on what day the shadow at sunrise points the farthest north or south. And there are many constructions, including many ancient ones, which are aligned to the solstice.

Possibly the oldest one, dated to the Neolithic Age, that is, about 7,000 years ago—3,000 years earlier than the oldest Egyptian pyramid—is at Nabta Playa in the Nubian desert in southern Egypt, about 500 miles south of Cairo. Here there is a ring of stones and four sets of upright slabs. One pair of slabs seems to point toward the summer solstice. In Australia there is an egg-shaped arrangement of stones built before the arrival of Europeans that indicates the position of the equinox and the solstices. There may be solstitial alignments in the line of stones at Bighorn Medicine Wheel in Wyoming and at the Cahokia Mounds near Collinsville, Illinois. On the island of Tonga, a trilithon—two large upright stones with a third stone across the top—constructed of coral slabs seems to point to the winter solstice. On Moloka'i in the Hawaiian Islands, a phallic rock is part of a solstice alignment. At Fajada Butte, the most conspicuous landmark in Chaco Canyon, New Mexico, rays of the midday Sun at the summer solstice cross a spiral etched into the rocky surface. A solstitial alignment consisting of two upright stones and a cobble-floor observing platform at Brainport Bay, Scotland, dates to the Bronze Age. And many more could be listed.

One of particular note is at Newgrange, Ireland. Here there is a prehistoric mound, built around 3000 B.C.E., which makes it older than the Egyptian pyramids, that was constructed of alternating layers of earth and stone. Built into the mound is a long passage that ends at a small room. Sometime after construction, when the bones of many bodies had been placed in the room, a large stone was placed at the entrance to the passage. Yet light could still enter the room through a long shaft. What is amazing is that the shaft is aligned so that direct sunlight can enter and illuminate the central room only on a few days at sunrise around winter solstice. What is more amazing is that within the small room is a rock slab with several symbols etched into it. One of those symbols might depict a solar eclipse.

The rock slab in question is known as "Cairn L" and the symbols carved on it are a series of concentric circles of different sizes. The largest two concentric circles overlap. It is these that are suggested to represent a solar eclipse.

Whether or not a solar eclipse is depicted at Newgrange is open to debate. No one knows for sure exactly what the concentric circles represent. But this much is clear: The builders of the mound at Newgrange understood the periodic motion of the Sun. What about the Moon?

There is a ring of stones around the mound at Newgrange. And there are other mounds nearby that have upright stones and interior shafts. Because there are so many possible pairings of stones with stones or with shafts, inevitably, some will align with lunar positions. But which positions are the most important?

Each month there is a northernmost moonrise and a southernmost one, analogous to the solstitial positions of the Sun, but from month to month those extremes in lunar position change. When watched over periods of years, the changes are substantial, so that those who watch the Moon refer to a "northern moonrise at major standstill" and a "northern moonrise at minor standstill," and corresponding ones for southern moonrise. All this leads to an important point: the Moon's motion is much more complicated than the Sun's motions. But that has not stopped people from trying to identify lunar alignments at ancient structures.

And when they do, the inevitable suggestion is made. If ancient people could follow the motions of the Sun and the Moon, then they might have been able to predict eclipses. It is a claim that has been most emphatically made at one of the world's most famous ancient structures, the ring of colossal stones at Stonehenge in southern England.

And yet the claim of eclipse prediction at Stonehenge is almost certainly wrong.

In 1961 Gerald Hawkins, a young astronomer from Boston University, was visiting Stonehenge to photograph the famous sunrise that one could see down the broad causeway on the morning of the summer solstice. As he stood there, he looked around at the many giant standing stones, some connected by lintels to form simple arches, and wondered if they might be marking other predictable events in the sky. When he returned home, he found a published map of the site. He drew lines that connected prominent stones and other features and that ran through the arches. Then, using a computer, which was still a novelty at the time, he computed where the Sun and the Moon would be seen to rise or set at Stonehenge at particularly notable times of the years, such as during an equinox or solstice. He compared those positions on the horizon with the lines he had drawn on the map and decided that there would about two dozen significant alignments.

In October 1963 he published his findings in the prestigious science journal *Nature*. A year later a television documentary was made about Hawkins and his work about Stonehenge. It emphasized that, according to Hawkins, if one stood within the great monument on the day of the summer solstice, June 21, one would see the Sun rise directly over a large upright stone known as "the Heel Stone." The final scene showed this was true. The documentary won a Peabody Award as "the most inventive art documentary of the year." Astronomers praised Hawkins for his imaginative work. The public's interest in Stonehenge soared. And then Hawkins took the next logical step.

He collaborated with Fred Hoyle, Britain's most famous astronomer, and the two men proposed how Stonehenge could have been used to predict eclipses. Archaeologists familiar with Stonehenge listened to their suggestions and considered Hawkins's other work. They were incredulous at what their heard. And they shook their heads "no."

Stonehenge "stands as lonely in history as it does on the great plain," wrote the American novelist Henry James when he visited the site in 1875. And that is part of the great mystery. Why do these colossal multi-ton stones, some transported hundreds of miles, lie on an otherwise featureless plain near the city of Salisbury?

The most striking feature is of course the great stones. Most are arranged in circular or horseshoe patterns. Several smaller upright stones lie outside the rings of large stones, notably the four Station Stones that form an elongated rectangle. There is also an outside ring of chalk-filled holes. These are the Aubrey holes, named for John Aubrey, who discovered them during an exploration of the site in the mid-1600s. The holes number fifty-six and are almost evenly spaced. They are the feature used by Hawkins and Hoyle to predict eclipses. And then there is the prominent Heel Stone that lies outside the ring of Aubrey holes and is slightly off the central axis of a broad causeway runs radially to what seems to be the center of the monument and out to the northeast.

The main objection archaeologists had about the ideas proposed by Hawkins and Hoyle is this: Stonehenge is not a single structure, but a series of structures that were built, modified, abandoned, rebuilt, and added to over a period of nearly two thousand years. At no time other than the present has Stonehenge ever looked the way it does today. Two stones that are close together today or that seem to be aligned may still have been erected hundreds of years apart.

The first suggestion that Stonehenge might have something to do with what is happening in the sky came from William Stukeley, a collector and dealer in antiquities, who, in 1740, published the book *Stonehenge: A Temple Restor'd to the British Druids*. Obviously from the title Stukeley was one who advocated the idea that Stonehenge had been constructed by the Druids, a religious sect of Celtic people who date from around the second century B.C.E. Stonehenge is much older than that. Its oldest features, such as the Aubrey holes, date from around 3100 B.C.E., which is several hundred years before the construction of the great pyramids of Giza in Egypt.

Thirty years after Stukeley published his book, John Smith, a physician and a self-described "inoculator of smallpox," arrived at a parish about six miles south of Stonehenge and began his practice. At first he had the support of the local people, but that changed quickly and he was forced to abandon his practice when, in his own words, "every act of violence" was brought against him. To divert himself from his troubles, he directed his attention on Stonehenge, deciding, after a casual inspection of the site, as he put it, "without an Instrument or any Assistance whatever," he concluded that the key to understanding the monument was the location of the Heel Stone along the causeway in such a manner that "the apex of the stone" points directly toward the summer solstice. Ever since that pronouncement, almost everyone who has written about Stonehenge has repeated Smith's contention. That is, almost everyone except archaeologists who began a careful survey and an excavation of Stonehenge in the early twentieth century.

After decades of work, they concluded that the earliest features constructed at Stonehenge were the ring of Aubrey holes surrounded by a low embankment and a shallow ditch outside the embankment. Next, several of the Aubrey holes were covered when the four Station Stones and the Heel Stone were erected. Then the first ring of large stones was erected. The causeway and additional interior rings of stones followed.

The important point is that none of the Aubrey holes were exposed once the Heel Stone was erected. And so the Aubrey holes and the Heel Stone could not be used simultaneously to follow celestial motions or to predict eclipses.

Furthermore, the Heel Stone had a companion. There once was a large stone standing nearby, north of it along the causeway, that has since fallen. Its former presence was revealed in 1979 during an excavation that uncovered rubble in what was certainly a former hole. And this means that the Heel Stone itself and its location may not be an important marker—if it is a marker of anything.

The question must also be raised as to how precisely one can predict where the Sun will rise along the horizon at solstice. If one is standing close to the equator, it is relatively simple because the Sun rises at a steep angle—which is why tropical twilights are so short. At higher latitudes, as in the British Isles, the Sun rises at a shallow angle, the disk seeming to skim along a distant horizon. For that reason it is difficult to predict precisely where the Sun will first appear because the atmosphere distorts and shifts its location.

Everyone is familiar with looking at a glass of water with a straw in it. Where the straw enters the water, its location seems to shift. The same happens when looking through the atmosphere: The location of every object seen in the heavens is shifted upward from where it would be if the atmosphere was not present. It is a phenomenon known as atmospheric refraction and it is greatest near the horizon. In fact, near the horizon it is about half an angular degree (about 0.5°), that is, about the diameter of the Sun, which means, when you see the Sun fully risen with the lower limb on the horizon, the *actual* Sun is still below the horizon.

The problem is that the amount of atmospheric refraction near the horizon can vary greatly. Refraction is the reason distant objects seem to shimmer and why they may have unusual shapes or colors. It can create illusions such as mirages. The amount of atmospheric refraction depends on the temperature structure of the atmosphere and whether it has a high content of water vapor. For these reasons, even today, it may be impossible to predict the moment of sunrise to better than a few minutes or to predict where the Sun will rise along the horizon to less than one degree (1°), which is much greater than the precision of the stone alignments Hawkins has proposed at Stonehenge.

All in all, this means that the alignments proposed at Stonehenge or elsewhere are not precise enough to predict the motions of the Sun or the Moon and are certainly not sufficient to predict eclipses. But that does not mean that these ancient structures have nothing to do with the sky. They do have elements, such as the general alignment of the broad causeway at Stonehenge toward the summer solstice, that have a symbolic, rather than a predictive, purpose.

The fact that our distant ancestors did watch the sky, and did so carefully, including the recording of eclipses, is all too apparent for a variety of reasons. As already mentioned, there is the possible recording of a solar eclipse etched in a stone slab at the end of the long shaft at Newgrange. One of the oldest depictions of the sky is on a bronze disk known as the Nebra sky disk. It was discovered in Germany and shows in gold inlay a large disk and a large crescent surrounded by small disks. This too might record a solar eclipse when planets and bright stars became visible. There are also possible references to an eclipses in numerous oral traditions, including those that are familiar to many people today. In the book of Genesis, in the story of Abraham in Canaan, is written: "And when the Sun was going down . . . a great darkness fell upon him." In the *Odyssey*, after the suitors of Penelope have sat down to a midday meal, the seer Telemachus stands and addresses them, noting that the walls have turned dark and the suitors are shrouded as if by night from their heads to their knees as "the sun is blotted out of the sky." These are all clearly references to eclipses, but there is no indication that anyone had yet made an accurate eclipse prediction.

Then what to make of the proposal by Hawkins and Hoyle that Stonehenge might be an ancient calculator that could predict eclipses?

They based their claim on the fifty-six Aubury circles. They suggested that four markers, perhaps large stones, could have been moved around the ring of circles. One marker each represented the Sun and the Moon, and the remaining two, reminiscent of Rahu and Ketu, the lunar nodes. According to the rules Hawkins and Hoyle proposed, the Sun marker would be moved alternatively six markers one week and seven the next week. The Moon marker would be moved two Aubrey holes every day. And the markers that represented the lunar nodes would be moved three holes each year.

Could this have predicted an eclipse? Maybe. It would have predicted the *possibility* of an eclipse, that is, it would indicate when the Sun and the Moon were near a node, but whether any particular eclipse would be visible from Stonehenge cannot be decided by this technique. Furthermore, as Hawkins and Hoyle admitted, eventually, the markers would be out of position with how the Sun, the Moon, and the nodes were positioned in the sky, and the markers would have to be reset.

And so it seems too complicated and too arbitrary. Furthermore, there are other henges—monuments that resemble Stonehenge in its early stage before it acquired its distinctive circles of massive stones—where rings of chalk-filled holes have been found, though the number of holes vary from as few as five to as many as a few dozen. Only Stonehenge has exactly fifty-six. And so these other rings must have been for another purpose. Or, what seems more likely, the Aubrey holes have nothing to do with eclipse prediction.

When Hawkins and Hoyle made their proposal, it was the early days of a burgeoning new field now known as archaeoastronomy, a field that attempts to understand cultural traditions and the meaning of ancient structures by combining techniques already in use by archaeologists, anthropologists, and astronomers. Hawkins, with his brash original proposal about Stonehenge, though now proven untrue, did accomplish something important: He managed to attract a great deal of professional interest to the field.

But the story of how people of the Stone Age and the Bronze Age tried to comprehend and follow eclipses does not end there. Some of them may have had a crude way to predict the *possibility* of an eclipse using a technique far easier than the one proposed by Hawkins and Hoyle, and that did not require construction of huge stone structures.

All it required was to count the number of full moons.

––––––––

On the rocky desert of northeastern Mexico, near the small village of Presa de la Mula in the state of Nuevo León, there is a place where ancient hunters once stood. We know this because many spearheads and other artifacts can still be found alongside numerous carvings on nearby rocks that depict different aspects of hunting.

There are carvings that show human figures holding spears. Then there are shapes that appear to be stylized deer antlers or hoof prints. But on one of the rock faces there is something decidedly different.

Just down below a rock crest a large grid has been drawn and within each cell of the grid is a set of short vertical marks. Standing in front of it—the grid measures about ten feet across and three feet high—one feels compelled to count the number of marks. There are 207 of them. That number could be important to an ancient hunter because it is the number of days in the gestation period of the female whitetail deer that the hunter often pursued. But it is also the number of days it takes for seven full moons to appear in the sky.

As one continues to stare at the grid, as William Breen Murray, an American anthropologist and professor at the nearby University of Monterrey, has done many times during the last forty years, one sees another way to tally the marks. And a different number is produced: 177. Murray has found the same numbers represented elsewhere at Presa de la Mula, as well as other sites of petroglyphs in northeastern Mexico, usually occurring as tight clusters of dots pecked into the rock. Murray knows the significance of 177: This is the number of days that must pass to see six full moons.

No one knows when the rock carvings at Presa de la Mula were made, but they probably predate the Maya who lived to the south by thousands of years. First, the carvings are primitive in form, that is, there is no elaboration of human forms. There are no symbols used in Mayan writings or on their architecture. And many of the carvings overlap earlier ones, suggesting that they were made over a long period of time. And Presa de la Mula is not an isolated site. There are more than six hundred known petroglyph fields that follow a corridor through northeastern Mexico. It is also known that human occupation in the area goes back at least 8,000 years. And so the tally marks and the clusters of dots at Presa de la Mula and elsewhere were almost certainly done by people thousands of years ago. But could people, who were possibly nomadic and who had only simple implements, such as rock scrapers, watch the sky and actually predict eclipses? Yes.

Imagine you are standing somewhere in the western United States on April 14, 2014, and you look up at the night sky. You see a full moon turn a dark copper color, a total lunar eclipse. You decide to keep a tally of the number of days until you see another one. You wander around the western United States and, 177 days later, on October 8, 2014, you see another total lunar eclipse. You wander about and watch again. Another eclipse occurs 178 days later at moonset on April 4, 2015. And another one, this time at moonrise, occurs 176 days later on September 27 of the same year.

You wonder if, by counting six full moons, or about 177 days, you might be able to predict lunar eclipses. And so you decide to test the idea.

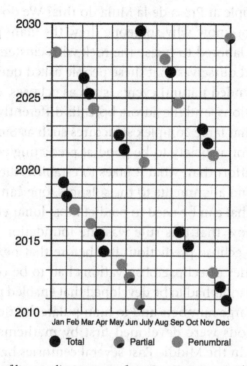

The occurrence of lunar eclipses as seen from San Francisco in the western United States from 2010 to 2030. The eclipses do not occur randomly, but in distinct bands that are separated by about 177 days.

The next total lunar eclipse visible from the western United States is on January 31, 2018. Starting on that date, you count six full moons. It is July 27, and you do *not* see one. But you are persistent and you count *another* six full moons, and, this time, on January 20 of the next year, there *is* a total lunar eclipse.*

You will have learned much in these twelve lunar cycles. By doing nothing more than watching the night sky and keeping a count of

_____

* A total lunar eclipse will occur six months earlier on July 27, 2017, visible across Asia, Africa and South America, but not from North America.

the number of new moons by tallying the number of days, you determined that, if you see a total lunar eclipse, it is highly *possible* that another will occur about 177 days later.

Did the people at Presa de la Mula do this? We do not know. If fact, we do not know why someone drew the giant grid at Presa de la Mula or labored to incise the rock with clusters of dots. We should remind ourselves that these people asked questions about and comprehended natural events, such as eclipses, earthquakes, volcanic eruptions, and the aurora borealis differently than we do. The point is that large complex structures such as Stonehenge are not required for a society to be good at predicting eclipses. With a little more effort than what it takes to establish the direction to the equinox and alignments to the solstices, one can discover the 177-day rule that can be used to predict most lunar eclipses.

And we know that this rule was the foundation for the next major step in eclipse prediction. But before that next step could happen, decades of eclipse observations had to be compiled and mathematical tools had to be developed that enabled people to add and subtract large numbers and to manipulate fractions.

Such methods were developed first by mathematicians who were working in the Middle East several centuries before the start of the Common Era. It was these people who would make the first accurate eclipse predictions.

CHAPTER THREE

Saros and the Substitute King

*The Sun will be eclipsed on the twenty-ninth of Iyyhar
. . . there will be a revolt in Akkad: son will slay his
father, brother will slay his brother (and) the king
will die and there will be fighting in the temple of Bel.*

—Babylonian eclipse prediction,
c. fifth century B.C.E.

S ome of the most memorable characters in world literature
make their first appearance in the vast collection of folk
stories known as *One Thousand and One Nights*. There is Ali

Baba and the Forty Thieves, Aladdin and his magic lamp, and the adventurous sailor Sinbad. One of the lesser-known of these characters is a ne'er-do-well who lives in the backstreets of Babylon named Abu al-Hasan.

One morning, as told in the story "The Dreamer and the Waker," al-Hasan is surprised to wake up and find himself in a royal bedroom surrounded by servants. They cater to his every need. They feed him the finest food and dress him in exquisite robes. All the while, unknown to al-Hasan, the real king is watching him from a hiding place, and, knowing the peril that al-Hasan faces, the king laughs.

The remainder of the story will remain untold so that it is not ruined for those who have not yet read it. But it is important to point out that for centuries this story produced a great deal of discussion among scholars who wondered if the story of Abu al-Hasan—the substitution of a commoner for a king—might be true. That is, was it the practice of some ancient civilizations to substitute a commoner for a king to protect the king from some dire fate, say, from the consequences of a bad omen? Perhaps due to an eclipse? There is an example of this, involving one of the towering figures in ancient history: Alexander the Great.

As it has been pieced together from several Greek authors, each of whom seemed to know only part of the story, it was the spring of 323 B.C.E., and Alexander the Great had just returned to Babylon after a long campaign in India. As he approached the city, several priests tried to dissuade him from entering, possibly because three eclipses were soon to occur in rapid succession, two partial solar eclipses on either side of a four-week period and a partial lunar eclipse midway between them. Exactly how the priests could have made such a prediction will wait until later. Alexander, for his part, ignored the warnings and entered Babylon anyway.

When he reached the palace and was ready to take his place on the throne he saw that there was already a man sitting there. Some accounts say that the man was a convict recently taken out of jail. Nevertheless, he was dressed in Alexander's royal robes and was wearing a crown. There was also a small army of palace eunuchs surrounding him and protecting him from an attempt by anyone who might want to remove him from the throne. Time passes—the accounts are not specific as to how much—and the priests advise Alexander to execute the man, and Alexander orders it done.

If this was a ritual to protect Alexander from an early death—he was thirty-three years old at the time—it did not have the intended effect. Soon after the three eclipses occurred and the man was executed, Alexander contracted a high fever. He died ten days later.

And this is where the question of the substitute king ritual stood, that is, until 1957 when clear evidence was finally found that the ritual was not only practiced, but that it was used frequently.

The discovery was made by Wilfred George Lambert, a specialist in the archaeology of the Near East, who, for more than a decade, had pored over the thousands of broken and cracked clay tablets with cuneiform text that were taken decades earlier from the ancient city of Nineveh and that now reside at the British Museum in London. After years of transcribing, deciphering, translating, and cataloguing a small segment of this vast collection, he realized that there were three fragments that, when put together, gave definitive proof that a ritual involving a substitute king had been performed—and that the performance required the prediction of an eclipse.

The three fragments constituted a letter from an astrologer to Esarhaddon, the king of Assyria and Babylon. The letter states that a total lunar eclipse is to occur on the date that we know as November 22, 671 B.C.E. It identifies the man who will be the substitute king as "Damqî." It even gives the name of the prophetess, Mullissu-abu-usri, who took the king's clothes to dress Damqî.

The substitute king was treated with honor. He participated in rites of ablution and purification, he recited litanies, and, later, after the eclipse had passed, he was put to death. The fragments describe the rites that followed and how the funerary chamber was prepared.

The true king, in the meantime, had to behave as inconspicuously as possible. He avoided being seen, but when he was, he was addressed as "farmer" or "peasant." After the death of the substitute, the real king resumed his role.

Other tablets show that this was not the only time a substitute king was used. During his twelve-year reign, Esarhaddon used it as many as eight times. And every single performance of the ritual during the reign of Esarhaddon can be traced back to an eclipse, either a lunar or a solar one.

But how did the small army of people—the scribes, the astrologers, the diviners, those who arranged the rituals, the healers, and the lamentation priests—who were advising Esarhaddon become so good at predicting eclipses?

It began with the 177-day rule and was expanded after a much longer eclipse cycle was discovered.

————•————

After watching a series of lunar eclipses occur at 177-day intervals, that is, at every sixth full moon, one occasionally finds that this is briefly interrupted when a lunar eclipse happens after only 148 days or only five full moons. But a new rule can be added, one that says: If you see a lunar eclipse, the next one will happen in 148 days *or* 177 days. By doing so, the ability to predict a lunar eclipse rises to about 75 percent, quite a respectable percentage, and, yet, one would like to do better. One wants to be able to predict *all* eclipses, and at what time of day or night one will occur, and in what part of the sky.

The immediate problem with this is the same as the problem one has in trying to keep a calendar synchronized with the movements of both the Sun and the Moon. The problem is commensurability.

Two numbers are commensurable if they can be expressed as a ratio of two integers. For example, if it took *exactly* 360 days for the Earth to go around the Sun and *exactly* 30 days for the Moon to go around the Earth, then it would be easy to construct a calendar. There would be twelve months of exactly 30 days and one could decide that the first day of every year would have a full moon, which means the first day of every month would also have a full moon.

This would repeat year after year, and it would be so easy to keep track of where the Sun and the Moon are in the sky. But the time it takes the Earth to orbit the Sun and the time it takes the Moon to orbit the Earth are not these simple values.

Instead, the time it takes the Earth to orbit the Sun—known as the *tropical year*—is 365 days, 5 hours, 49 minutes, or in decimal form, 365.2422 days. And the time it takes the Moon to go from full moon to full moon—the *synodic month*—is 29 days, 12 hours, 44 minutes, or 29.5306 days. If we divide these two numbers, we discover that, after one year, the Moon has orbited the Earth twelve-and-a-fraction (more precisely, 12.235) times. Moreover, the length of a tropical year and the length of a synodic month are not constant; they vary slightly due to a variety of sources, such as the gravitational effect of Venus on the motions of both the Earth and the Moon. A tropical year and a synodic month are not commensurable; and so it is impossible to construct a yearlong calendar that keeps the beginning of a synodic month in sync with the beginning of a tropical year.* The same problem occurs when one tries to predict eclipses.

If the time it took the Moon to go from full moon to full moon was *exactly* thirty days and if the time it took the Moon to go from one lunar node through its orbit and back to the same node was *exactly* twenty-eight days, then it would be easy to predict eclipses because it would take only 420 days to see every possible configuration of the Moon relative to the lunar nodes in the sky.** That means the series of eclipses that occur in one 420-day period will

---

* However, one can get close if one is willing to have a calendar that covers multiple years. Notice that 19 tropical years is almost equal to 235 synodic months, that is, (19 × 365.2422 days = 6939.602 days) and (235 × 29.5306 days = 6939.691 days). So that after 19 years, the beginning of a tropical year and a synodic month will be out of sync by only 0.089 days, or about two hours. This 19-year cycle is known as the Metonic cycle. It was known to the Babylonians and is the basis of some modern-day calendars, such as the Jewish calendar.

** Because (14 × 30 days = 15 × 28 days = 420 days), the relative positions of the Moon and of the lunar nodes would repeat *exactly* every 420 days.

reoccur *exactly* in the next 420-day period, and so forth. You would only have to observe the sky for 420 days to predict when all future eclipses will occur.

But the motion of the Moon is not that regular. The synodic month of 29.5306 days is not commensurable with the draconic month—the time it takes the Moon to move from one lunar node and return to the same node—of 27.2122 days. However, after many years, a long series of synodic months *almost* becomes synchronized with a long series of draconic months. Specifically, 223 synodic months equals 6,585.3238 days and 242 draconic months equals 6,585.3523 days. The difference, after more than 18 years has passed, is only 0.0285 days, or 41 minutes.

And there is a bonus. The path the Moon follows around the Earth is not a circle, but an ellipse. As it follows this elliptical path, two important things happen: the distance between the Earth and the Moon changes and the speed that the Moon moves across the sky changes. The time it takes for the Moon to go from the most distant point of its orbit, known as the *apogee*, circle the Earth, and return to the same point is the anomalistic month, and its duration is 27.2122 days. And 239 *anomalistic months* is equal to 6585.4538 days, about three hours difference from 223 synodic months.

What does this mean? If you find a series of eclipses, that series will repeat 6,585 days later, that is, 18 years, 10 or 11 days later. (Whether it is 10 or 11 days depends on how many leap days have occurred.) For example, in 1998 the eclipse series was solar-lunar-solar-lunar with events occurring on February 26, March 13, August 22, and September 6. Eighteen years later, the same series of solar-lunar-solar-lunar events occurred, shifted by 10 or 11 days, on March 9, March 23, September 1, and September 16.

It also means that the detailed geometry of eclipses separated by this interval is almost the same. This is best illustrated by comparing the totality paths of solar eclipses.

The path of the total solar eclipse on August 21, 2017, will be in the northern latitudes crossing from northwest to southeast across the United States. Eighteen years and ten days earlier, on August 11, the path of a solar eclipse crossed Europe from northwest to southeast. And on September 2, 2035, the path will be across Asia and the western Pacific from northwest to southeast.

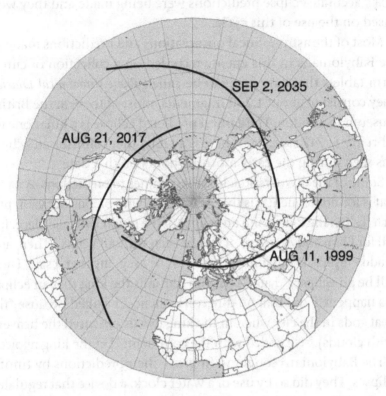

The similar ground tracks of three solar eclipses separated by one Saros cycle. Each track is offset by about 120° in longitude.

The cycle of eighteen years and ten or eleven days has been so important in the history of predicting eclipses that it has been given a name. It is known as the *Saros cycle*, named by Edmond Halley in 1691—who is best known for the comet that bears his name—who rediscovered the cycle after it had been forgotten for more than a millennium by searching through eclipse records in ancient Babylonian texts.

When did the Babylonians first discover the Saros cycle? That remains unknown because no texts have yet been found that could answer the question. Instead, by the middle of the eighth century B.C.E., accurate eclipse predictions were being made and they were based on the use of this cycle.

Most of the astronomical observations and predictions made by the Babylonians in this era are recorded on a collection of cuneiform tablets that have come to be called *The Astronomical Diaries*. They consist of about 1,200 fragments, most of them at the British Museum in London. The earliest predicted eclipse is a lunar one for February 6, 747 B.C.E. And there is a tablet that confirms an eclipse was seen during the early morning hours on that date.

Some of the diviners attached their names to predictions. A name that is found frequently is that of Irassi-ilu who on one occasion predicted: "On the fourteenth an eclipse of the moon will take place. It is evil for Elam and Amurru.* It will be seen without Venus." Then, as if to add emphasis, Irassi-ilu writes: "To the king, my lord, I say, 'There will be an eclipse.'" Later, Irassi-ilu reminds the king that an eclipse has happened and that no dire consequences resulted because "the great gods in the city wherein the king dwells obscured the heavens (with clouds)." He ended with the comment: "Let the king rejoice."

The Babylonians sought to improve their predictions by timing eclipses. They did so by use of a water clock, a device that regulated the flow of water from one large vessel to another. Because such a clock could run for only a few hours, if a lunar eclipse was expected for the first half of a night, the Babylonians would start a water clock at sunset and read off the elapsed time at the beginning of the eclipse. If an eclipse was expected to occur after midnight, they would measure the time interval between the eclipse and sunrise.

From the cuneiform texts we learn that eclipses were computed by simple arithmetic. The calculations are arranged in columns. An indication that motion of the Moon was by far the most advanced

---

* Elam and Amurru are places near the city of Babylon. Elam is to the east and Amurru is to the west.

part of Babylonian astronomy is that the calculation of planetary motions, for example, determining when Venus would rise required four or five columns, while a calculation of the position of the Moon could run to eighteen columns.

The Babylonians predicted both lunar and solar eclipses, though their lunar predictions were far more accurate. The beginning of a lunar eclipse could usually be predicted to within two hours. The timing of a solar eclipse could be missed by several hours. The difference is easy to understand. During a lunar eclipse the entire Moon is covered by the Earth's shadow, while during a solar eclipse the Moon's shadow falls only on a part of the Earth. This means an accurate prediction of a solar eclipse must take into account an observer's position on the Earth's surface. There is no evidence that the Babylonians ever considered this factor.

Also they had many more observations of lunar eclipses than solar ones simply because lunar eclipses are easier to see. It is difficult to notice a solar eclipse with the naked eye unless the Moon obscures nearly 90 percent of the Sun's disk. And so they were limited in what they could accomplish with regard to solar eclipses.

But there was another ancient civilization equally interested in eclipses. And through celestial happenstance they were able to predict solar eclipses with an astounding degree of accuracy for the first time in history.

One of the dark days of history is July 12, 1562. On that date Spanish friar Diego de Landa had thousands of documents kept by the local Maya brought into the main square of the Yucatán town of Maní and burned. He was following a long-established practice. In 1529 Juan de Zumárraga, the first Bishop of what was then known as New Spain, had the contents of an Aztec library emptied and, as witnesses recorded it, piled into "a mountain heap" in the town square. Then monks paraded by carrying torches and singing and setting fire to the heap. In 1499 the Archbishop of Toledo in Spain,

Jiménez de Cisneros, convinced Muslims to bring out books in Arabic. He then ordered the books thrown into a bonfire.

We are much the poorer for such willful actions. Moreover, we do not know exactly what was destroyed. As de Landa wrote: "We found many books written in small strange characters and because they contained nothing that was free from superstition or lies of the devil, we burned them all." He did count these books, or codices. There were twenty-seven of them. He also noted that the Maya "regretted to an amazing degree" the destruction and that it "caused them much affliction."

Only three Mayan codices and the fragment of a fourth survive today. Presumably, they were seized by Spanish conquerors and taken back to Spain before they, too, were consigned to a fire. The three codices are now scattered across Europe. One is in Madrid, another in Paris, and the third in Dresden, Germany. By far, the largest and the best preserved is the Dresden codex.

The first account of the existence of the Dresden Codex is in 1739 when Johann Goetze, director of the royal library at the court of Saxony, purchased it from a private library in Vienna. Its earlier history is unknown.

The codex consists of seventy-four pages on thirty-nine double-sided sheets. It is eight inches high and overall has a length of nearly twelve feet and folds together like an accordion. The pages are made from thin sheets of bark that has been covered with white lime. The symbols are painted with vegetable dyes. Four pages are blank. The remaining pages are covered with an amazing array of symbols and small drawings. Most pages are dominated by small figures, known as *glyphs*, arranged in rows and columns. Each one has been carefully painted on with a fine brush in reds and blacks and blues. There is usually a column or two of dots and horizontal lines and the occasional large drawing, often showing a half-human, half-animal creature wearing an elaborate headdress. Close examination of the glyphs indicates that they were written by at least four different scribes. There are also indications that the Dresden Codex is a copy of an earlier document. As to when the codex itself was made, opinions differ greatly, ranging from 1200 to 1519 C.E. The latter date is the year of the Spanish conquest.

Not until 1853 was the Dresden Codex recognized to be a Mayan manuscript. In 1880 Ernst Förstemann, who was then the chief librarian in Dresden, realized that the codex contained several references to calendars and to the planet Venus. Others noticed that a large section of the codex had something to do with eclipses.

The idea that eclipses are part of the Dresden Codex came from studying the series of dots and horizontal bars that appear on many of the pages. These dots and bars are the way that the Maya expressed numbers. And they are readily decipherable.

For example, on pages 52 and 53 of the codex, the following numbers are listed: 6,408, 6,585, 6,762, 6,939, 7,116, and 7,264. (The astute reader will already recognize "6,585" as the number of days in a Saros cycle, a quick indication that this sequence of numbers probably has something to do with eclipses.) If one takes the differences between successive numbers, the following is obtained: 177, 177, 177, 177, and 148. This sequence of numbers is also written at the bottoms of pages 52 and 53 of the codex. And those two numbers, 177 and 148, should be instantly recognized as the number of days that pass for one to see six or five full Moons, an interval easily determined between successive eclipses.

At the end of what seems to be an eclipse section one finds the number 11,959, a curious choice until one realizes that this is one day short of the number of days in 405 synodic months, a further suggestion that this section of the Dresden Codex is somehow related to the movement of the Moon. But what is actually represented?

Researchers have studied this amazing document. Some have concluded that it is a record of eclipse observations. But no one has yet been able to show how the pattern of eclipses in the Dresden Codex would correspond to an actual sequence of eclipses that would have been seen by the Maya. Instead, it is more likely that it is an eclipse *warning* table, one that would alert the Mayas when an eclipse was possible.

Of course, to make an eclipse-warning table it was necessary to compile many observations of eclipses. Unfortunately, how the Mayas may have produced the contents of the Dresden Codex is lost, possibly consumed in the fire set by Diego de Landa or by other acts of destruction. But we do know how one could have been made.* And we know that the Dresden Codex *could* have been used to predict *both* lunar and solar eclipses. In fact, judging from other evidence—other written codices and artifacts—it seems that the Maya were even more interested in solar eclipses than the Babylonians were. Why?

A hint is given by looking at when solar eclipses seem to cluster. Yes, they do occur at regular intervals the same as lunar eclipses do. However, solar eclipses, which are usually partial eclipses, are seldom noticed because blinding light is still coming from the uncovered part of the Sun. (That is, unless the Sun is almost totally obscured.) And if there is a quick sequence, or a clustering, of such total or near-total solar eclipses, then there is probably a tendency to keep a closer lookout for any solar eclipse.

How often do solar eclipses cluster? Not very often.

The Belgian astronomer Jean Meeus who is one of today's foremost experts on modern eclipse calculations has shown that total solar eclipses take place at very irregular intervals for a given place. He was dismayed to discover that for his home city of Antwerp the next total solar eclipse will not take place until May 5, 2142, and that it would be the first such event in at least seven centuries. And yet then there would be a second total solar eclipse across Antwerp just nine years later on June 14, 2151.

He then started to look to see if total or near-total solar eclipses might cluster. He found three such events happening in south-central China during his lifetime, in 2009, 2010, and 2020. And that four occurred recently over Laos in just fourteen years between 1944 and 1958. And so it should be considered amazing that *five*

---

* Anthony Aveni, who has studied the Dresden Codex and many other aspects of Mayan culture, has laid out a way to make an eclipse table in his book *Stairways to the Stars*.

total or nearly total solar eclipses were visible from the Yucatán Peninsula during the thirteen years from 331 to 344 C.E. This was only a century or so after the beginning of what is regarded as the classic period of Mayan civilization, when the first sophisticated Mayan calendar appeared as well as the beginning of large-scale construction. It was also the period when, from temple inscriptions, Mayan astronomy began.

It is reasonable to propose that these five solar eclipses occurring over a period of barely more than a dozen years might have led to keen interest in watching and recording future solar eclipses, even those that only slightly obscured the Sun. And that could have led to the eclipse table given in the Dresden Codex.

Sadly, we may never know. Written evidence has not been found—and probably never will.

———•———

Mayan knowledge about eclipses can be found in the later Aztec civilization, but that is as far as it was transmitted, halted by the Spanish conquest. By contrast, what the Babylonians knew about eclipses was diffused widely.

It moved into India and China and then into Japan. It also moved to the West where one might expect the Egyptians would adopt it, but here is one of the oddities of ancient eclipse watching.

For all their grand temples covered with inscriptions, the ancient Egyptians were completely silent about transient astronomical phenomena such as eclipses and comets. From the almost three millennia of Egyptian writing, the only texts that have come down to us that mention eclipses are in the late dynasties that were heavily influenced by the Greeks and the Romans. The reason for the scarcity is not known. Perhaps an eclipse was such a bad omen that there was silent relief when one passed and it was thought better not to tempt fate and record it. Or the Egyptians may have recorded such events on papyrus that did not survive or that has not been found. Whatever the reason, though the Egyptians left no eclipse records and showed no signs of embracing Babylonian astronomy, those who lived to the north in the Greek islands, on the mainland, and in Asia Minor did.

The main transfer of astronomical knowledge from the Babylonians to the Greeks occurred during the fourth century B.C.E. during the time when the Greek philosopher Aristotle was tutoring Alexander the Great. As a result of the close relationship, Alexander ordered the translation of many Babylonian astronomical texts into Greek. A century later, the Greeks had improved on the Babylonian methods, reducing the error in timing eclipses to about thirty minutes.

By that time the Greeks had also invented a number of simple astronomical instruments. The *diopter* was an early land-surveying instrument that was also used to measure the angular distances between stars. The *tetrantas* was a primitive forerunner of the navigator's quadrant that could be used to determine latitude by pointing it at Polaris, the North Star. The astrolabe, also used to determine latitude, would go through centuries of development, but the earliest versions were invented by the Greeks to determine where the Sun or the Moon or various stars would rise or set. Given what seemed to be a limited ability to design and build astronomical equipment, might it be possible that the Greeks of this period could have built a sophisticated machine that would simulate the complex motions they saw in the heavens? The general opinion was "no." But then, a few years ago, a group of scientists reexamined a small mass of bronze machinery retrieved from the floor of the Mediterranean Sea more than a hundred years ago. And they came to a very different conclusion.

In 1900 a group of Greek sponge divers was returning home from the coast of North Africa. They were passing through the broad strait between the mainland of Greece and the island of Crete when a storm forced them to find shelter close to the small barren island of Antikythera. After the storm subsided they decided to explore the rocky shelves beneath them. At a depth of about 120 feet, they found the seafloor was scattered with bronze and marble statues.

They investigated further, discovering that what they had stumbled upon was an ancient shipwreck.

Over the next year more than a hundred statues were retrieved; some representing the finest workmanship ever discovered in Greece. Clay amphorae still containing wine and hundreds of coins were also found. A study of the coins showed that they originated east of the wreckage in Asia Minor and that they date from the first century B.C.E. Also found was a lump of bronze calcified by seawater and covered with seashells. It was taken to an archaeological museum in Athens where it remained in a crate until 1902 when it was finally examined. The first cursory examination revealed small, indecipherable Greek inscriptions. Further examinations showed that it had perfectly cut small triangular gear teeth.

Attempts to conserve the piece caused it to separate into three main fragments. Today it exists as seven large fragments and scores of small ones. The large fragments were again subjected to close study. This time evidence was seen that the bronze mass was once enclosed in a wooden box. There had been a large dial on one side and two smaller dials on the opposite side. The whole thing had once been about the size of a modern laptop computer or a shoebox. And inside the largest fragment was a complex mechanism composed of dozens of bronze gears.

In 2005 new techniques were applied to study the object, which by now was known as the Antikythera Mechanism. It was subjected to X-rays to produce a high-resolution, three-dimensional image of the hidden inside. That revealed a complex clockwork mechanism of interlocking gears. The texture of the outer surface was mapped in great detail by illuminating the mechanism at different angles by strong light. The result showed hundreds of Greek characters not yet seen, some less than a tenth of an inch in size.

Years were required to map the pattern of gears and to decipher the newly found Greek characters. One of the first major discoveries was that the largest gear had 223 teeth. Those who were studying the Antikythera Mechanism knew immediately what this meant: 223 is the number of months in a Saros cycle. And there was more.

The Antikythera Mechanism was a highly sophisticated machine that could calculate the cyclic motions of the Sun and the Moon and the planets by a rotation of gears. In short, it was a mechanized version of Babylonian and Greek astronomy. The main dial showed the positions of these celestial objects relative to the zodiac, that is, the seasonal patterns of stars. One of the dials on the opposite side kept track of solar and lunar motions over a repetitive nineteen-year period known as the Metonic cycle. The other dial, the one that the large 223-tooth gear was attached to, could be used to predict eclipses.

To make a prediction, the dial had to be synchronized with a lunar or a solar eclipse that had recently been observed. Then, by turning a handle on the side of the mechanism, the gears would turn and the dial would move. When the dial pointed to a particular inscription, the user of the Antikythera Mechanism would know when an eclipse was predicted, including the time of day and whether it would be partial or total.

How accurate was it? If it worked perfectly—for example, if the gears were perfectly spaced—then it could be used to predict a sequence of eclipses over centuries. If it was properly synchronized to recent lunar and solar eclipses, then an eclipse prediction could be made years in advance to an accuracy of a few hours.

But the gears, which are about a tenth of an inch in size, were not perfectly spaced because they were hand cut. And so an eclipse prediction might be correct to only a day or two and the location of the Moon in the sky predicted by the Antikythera Mechanism might be several lunar diameters from the actual position. And so Greek knowledge of solar and lunar motions, and their ability to predict eclipses through laborious calculations, was much better than what could be represented by the Antikythera Mechanism. Nevertheless, this machine was still an outstanding achievement. Nothing close to its sophistication is known anywhere in the world until the mechanized clocks of Medieval Europe.

In the subsequent centuries, progress was still being made on improving eclipse predictions, though not for the purpose of a substitute king, but for something far more practical.

CHAPTER FOUR

Measuring
the World

*When I was still a boy, I was called outdoors by my
parents to contemplate an eclipse.*

—Johannes Kepler, recounting a lunar eclipse
that he saw at age ten on January 31, 1580

The first map to show a significant part of the world was
drawn by the Greek geographer Ptolemy in the second
century C.E. It shows Spain in the west and what seems to
be the Malaysian Peninsula far to the east. The British Isles are
crudely drawn and what appears to be Denmark is nearby. The

most detailed part of the map, that is, the region where Ptolemy had the most information, was the Mediterranean. Here the distinctive boot-shape of the Italian Peninsula is clearly seen. The islands of Sicily, Sardinia and Corsica are shown in roughly the correct proportions, but the islands of Crete and Cyprus are much too large. But as one studies and contemplates Ptolemy's map, there seems to be something wrong with the general shape of the Mediterranean Sea. It seems too long for its width. The whole Mediterranean Sea and its many features are stretched out in an east-west direction. But Ptolemy drew it this way because, unbeknownst to him, a serious error had been made in timing a lunar eclipse.

Ptolemy and other ancients knew that lunar eclipses could be used to determine differences in longitude over great distances because such eclipses could be seen simultaneously over half of the Earth's surface, and thus the time and location in which they were viewed could be used to establish longitude. What was needed in order to do so were two observers to record where the Moon was in the sky during a particular phase of the eclipse, for example, the moment the Moon moved completely into the Earth's shadow. Then, by knowing the Earth's diameter—the Greek astronomer Hipparchus had determined a reasonably accurate value for the diameter just decades before Ptolemy drew his famous map—it was a matter of simple geometry to figure out the east-west, or longitudinal, distance between the two observers.

Ptolemy searched ancient eclipse records for two such simultaneous observations. He found a pair that had been observed about four hundred years earlier.

According to Babylonian records, as given in the *Astronomical Diaries*, eleven days before the Battle of Arbela when Alexander the Great defeated Darius III of Persia, a lunar eclipse occurred in the sky five hours after moonrise. Ptolemy was able to determine that the same eclipse was seen in Carthage in North Africa just two hours after moonrise. The difference of three hours meant the longitude difference between Arbela, which is close to the modern city of Erbil in Iraq, and Carthage was 45°. Actually, the time difference we now know should have been only two hours and the longitudinal difference only 30°. This error caused Ptolemy to overestimate the length of the Mediterranean Sea by fifty percent, about 600 miles.*

And it had historical consequences. Thirteen centuries later, when Columbus was trying to convince the King and the Queen of Spain to financially support his plan to sail west to China, he used Ptolemy's map—and its exaggerated east-west distances, though he did not know they were exaggerated—to show that China could not be as far away as others had estimated. In the end, Columbus did sail, and, thereby, history was changed, though not in the way he had intended.

The first planned attempt to use a lunar eclipse to determine a longitudinal difference over a long distance was done by the Muslim scholar Ahmad al-Biruni. He was in the city of Kath (near the modern city of Karakalpak, Uzbekistan) on the southern coast of the Aral Sea. He and a friend, Abu al-Wafa, who lived in Baghdad, agreed to make simultaneous observations by timing an eclipse. The lunar eclipse they chose was on May 24, 997. They determined that the difference in time of the eclipse was about one hour, a longitudinal difference of about 15°, close to the actual value.

Al-Biruni arranged for other simultaneous observations. In 1003 he determined the longitudinal difference between Baghdad and modern-day Ghaznai in Afghanistan to be 14°, which is the actual value; in 1004, he determined the difference to Kunya-Urgench in Turkmenistan to be 10°, one degree more than the actual value. Considering that he and his collaborators were timing the eclipses by using clocks that relied on the slow shifting of sand, forerunners of the marine sandglasses, which would be invented three hundred years later, to achieve such accuracy, they must have been able to determine local time to an accuracy of about four minutes over periods of several hours, a remarkable accomplishment, indeed.

---

* In comparison, latitudinal distances are easy to measure because a single observer can determine latitude by measuring the angular height of Polaris, the North Star, above the horizon.

Ptolemy flourished at the height of the Roman Empire. He lived in the city of Alexandria, and though born a Greek, was a Roman citizen. Beyond that, little else is known of his life, except that he managed to write several treatises. He wrote about optics and music and geography, using his knowledge of the latter to draw his famous map. But he was most prolific when writing about astronomy.

Originally given the Greek title *Mathematike Syntaxis*, "Mathematical Treatise," the original Greek versions are lost. It survives in its earliest form in Arabic. In fact, Arab astronomers were so impressed by its contents that they attached the superlative *megisti* to it. When the definite article *al-* was prefixed, Ptolemy's great book about astronomy became known in Arabic as *al-Majisti*. Later, when it was translated into Latin, it became *Almagest*, the title by which it is known today.

The *Almagest* is almost entirely concerned with the technical aspects of astronomical measurements and calculations, mostly based on techniques first developed by the Babylonians. Ptolemy has tabulated observations made by earlier astronomers. Whether he contributed any observations of his own is unknown. It relies heavily on the ideas of the Greek astronomer Hipparchus who lived four hundred years earlier. The *Almagest* is so comprehensive that it made all earlier astronomical writings obsolete, and so scribes stopped copying earlier works, which means Ptolemy's work is the only complete treatise on astronomy that survives from the ancient Mediterranean world.

The *Almagest* does have its problems however. For example, discussions about how to calculate and observe eclipses are scattered over several chapters. Fortunately, Ptolemy corrected this shortcoming and produced another work about astronomy, one that is widely known as the *Handy Tables*. In that book, he grouped similar types of calculations and illustrated them with extensive tables. It is a book that seems to be specifically designed for people who needed to do routine calculations, for example, astrologers who would be among the primary users. None of the pages of Ptolemy's original *Almagest* and only a few fragments of the *Handy Tables* survive. The earliest complete versions of either are copies made by the Greek mathematician Theon of Alexandria in the fourth century C.E.

A century after Theon made copies, the western half of the Roman Empire collapsed. The eastern half continued as the Byzantine Empire with its capital at Constantinople, the modern city of Istanbul in Turkey. There were almost constant military skirmishes between the Christian Byzantines and the Muslim Arabs to the east. Nevertheless, occasionally emissaries were exchanged. It was during one such exchange—one account puts it in the year 820—that an Arab scholar known to history only as "Salm" convinced the Byzantine Emperor Leo the Armenian to allow him to copy books in the library at Constantinople. Among the books he copied were the *Almagest* and the *Handy Tables*.

It was in this way that much of Greek knowledge of astronomy formally entered the Muslim world where it was greatly expanded upon. Foremost among them was Musa al-Khwarizmi, who combined what the ancient Greeks had known about astronomy with advancements that had been made by Indian mathematicians to improve the calculation of astronomical tables. The product was a collection of tables that predicted planetary movements, the time of sunrise and sunset, and when the lunar crescent would first be visible after a new moon. There was also a table that predicted eclipses. Al-Biruni may have used such a table to plan simultaneous observations of lunar eclipses.

By the start of the second millennium C.E. the center of Muslim astronomy had shifted from Baghdad in the East to the Spanish city of Toledo in the West. In 1080 another comprehensive set of astronomical tables were computed by a group of Muslim and Jewish astronomers and, because of where they were produced, are known as the *Toledan Tables*. Then, soon after the *Toledan Tables* were produced, much of what would become modern-day Spain was taken over by Christian rulers who, after a century of observations, determined that the predictions made by the *Toledan Tables* were often off by as much as a day. And so yet another new set of astronomical tables were computed.

These new tables were commissioned by Alfonso X of Castile, and so, are known today as the *Alfonsine Tables*. They were published in 1252 and, for some events, such as eclipses, gave *hourly* predictions, a revolutionary degree of accuracy for that era. They proved superior to the *Toledan Tables* in large part because Alfonso insisted that new observations be made. The whole enterprise of trying to predict celestial events, observe them, then correct the calculations, as well as employ the methods described by Ptolemy and improved by al-Khwarizmi, was a daunting one. And maybe this reason that Alfonso was prompted to say; "If the Lord Almighty had consulted me before embarking on creation, I should have recommended something simpler."

On reflection, a great deal of gratitude is owed to those Muslim astronomers who maintained and expanded humanity's knowledge about the sky. Christian Europe was then in a period of malaise—some have said it was a deterioration—known as the Dark Ages that followed the collapse of the Roman Empire. But once Europe started to rebound, the knowledge spread quickly. An indication of how quickly is provided by how soon Ptolemy's ideas appeared in Western literature.

The Italian poet Dante Alighieri was born thirteen years after the publication of the *Alfonsine Tables*. His greatest work, *The Divine Comedy*, written between 1308 and 1320, draws on the *Almagest*. In particular, in the third book of the *Divine Comedy*, "Paradiso," Beatrice guides the traveler Dante through nine celestial realms, each one corresponding to one of the nine concentric spheres that Ptolemy proposed surrounded the Earth.*

---

* In an earlier, lesser-known, work, *Convivio*, published in about 1305, Dante cited Ptolemy when he explains such astronomical features as the nature of the Milky Way, the variable brightness of Mars, and why Venus follows a complex motion in the sky.

A half-century after Dante, the English writer Geoffrey Chaucer demonstrated his considerable knowledge of Ptolemaic astronomy in *The Canterbury Tales*. In those tales, he mentioned the *Toledan Tables* once in "The Franklin's Tale" and the *Almagest* once in "The Miller's Tale" and twice in "The Wife of Bath's Prologue." Almost every one of the tales makes some reference to the movement or relative position of heavenly bodies. In "The Merchant's Tale" the blinding of the character Januaries by a scorpion is thought to be a reference to a lunar eclipse when the Moon was in the constellation Scorpius on May 10, 1389, an event predicted in the *Alfonsine Tables* and that Chaucer probably viewed.

The next major advance in producing astronomical tables came in 1474 when Johannes Müller, known to history by the Latin name "Regiomontanus," published a modification of the *Alfonsine Tables* using a printing press. His was an 896-page book, known simply as the *Ephemerides*, which sold quickly and was distributed widely. And there were many abbreviated versions of the *Alfonsine Tables* published. The one produced in 1496 by Abraham Zacuto in Portugal is of special note—his was titled *Almanach Perpetuum*, "Perpetual Almanac"—as Columbus took a copy on his fourth and final voyage to the New World.

The fourth voyage was his longest and most ambitious. He left Spain on May 12, 1502, and spent more than a year sailing along the coasts of Hispaniola, Cuba, and Central America. In June 1503, a storm damaged his flagship *La Capitana* and he decided to beach it at St. Ann's Bay on the north shore of Jamaica. He sent the other ships of his fleet to Hispaniola to arrange a rescue. Then he waited.

Months passed. The local people grew tired of the constant demands of Columbus and his men to supply them with food. Sensing the growing reluctance, Columbus sent a message inviting the local leaders to a meeting. They arrived on the evening of February 29, 1504, and stepped aboard the remains of the wrecked ship. Columbus informed them that God was angry with them for not bringing enough food to the ship. He said God would punish them through famine and pestilence. As proof of this prophecy, he predicted that the Moon would rise "inflamed with wrath."

Columbus made this bold assertion because he knew that there would be a lunar eclipse that night, having consulted the abbreviated tables produced by Zacuto. And the Moon did rise in eclipse. And the locals did respond by continuing to supply the demanded food. Finally, to everyone's relief, help arrived four months later on June 29, 1504, and Columbus and his men sailed away.

But it must be said that those with a whiggish sense of history will see Columbus's ability to predict a lunar eclipse and, hence, to survive, as the culmination of a series of events that led back to the *Alfonsine Tables* and al-Khwarizmi, to the unknown Arab scholar Salm, to Theon of Alexandria and Ptolemy and the *Almagest*, and finally to the ability of ancient Babylonians, fearing the death of their king, to predict eclipses.

———

Columbus's influence on the history of eclipses does not end with his extortion with an almanac to get more food. In 1494, two years after his first voyage to the New World, the Treaty of Tordesillas was signed between Portugal and Spain, the two principal nations then sending fleets of ships across the Atlantic Ocean. The purpose of the treaty was to define which part of the newly discovered land each country would control. The treaty specified that the division would be along a line of longitude, the one that passed 370 leagues (about 1,300 miles) west of the Cape Verde Islands off the coast of Africa. All non-Christian land east of the line would be under the political and economic control of Portugal, and everywhere west of the line would be controlled by Spain. The problem was: No one knew exactly which lands the line passed through. And so decades of disputes and, at times, armed conflict, ensued.

In 1518, when Ferdinand Magellan was preparing for a planned sail around the world, he decided he would help try to settle the matter by bringing with him an astronomer. He found one who drew up a list of lunar eclipses that might be seen during the voyage. Unfortunately, soon after the man went insane and never sailed with the expedition.

Magellan's voyage actually intensified the dispute. When his fleet returned to Spain—Magellan had died the previous year in the Philippines—the ships were filled with valuable spices obtained in the Maluku Islands located on the western side of the Pacific. But were these islands, which had a commodity that would obviously be important in future trading, under the control of Portugal or Spain? For the next several years, ships from the two countries engaged in an undeclared war in the Pacific. The matter was settled in 1529 by the Treaty of Zaragoza, which gave control of the islands to Portugal after a substantial payment was made to Spain. The treaty also defined a second longitudinal line, this one three hundred leagues east of the Malaku Islands, that set the division between Portuguese and Spanish influence. But, again, no one knew exactly which lands were on either side of the line. Spain decided to remove some of the uncertainty by setting into motion the first coordinated worldwide program to determine longitude, a program that would time lunar eclipses.

To do so, in 1577 the Spanish government created the new government post of cosmographer and chronicler major of the Council of the Indies. The first director of the program was Juan López de Velasco, a former legal assistant to the president of the Council of the Indies who ruled over all Spanish possessions. Velasco would run the program from Madrid for twenty years, the total lifetime of the program.

Though not a cartographer himself, Velasco surrounded himself with knowledgeable people who were. Then, in superb bureaucratic style, he had them produce a lengthy questionnaire to be sent out to government officials and major landowners. Answers to the questionnaire would be used to determine the location of provincial capitals and cities. It required estimating the distances and directions to major landmarks. It also required determining the angle of the Sun above the horizon at noon. Measurements were also to be made of the altitude of Polaris to determine latitudes. And, to complete the questionnaire, officials and landowners would have to observe lunar eclipses, the questionnaire giving detailed instructions on how to construct instruments and make the observations.

The first version of the questionnaire was prepared for the lunar eclipse of September 27, 1577, which would be visible from Europe and the Americas. More than a hundred questionnaires were printed and sent out. Only a few were returned. A second, shorter questionnaire with simpler instructions was sent out for a 1578 eclipse, and a third questionnaire, again with fewer questions and simplified instructions, was sent out for a 1581 event. Again, there were almost no responses. And those who did respond, their answers and measurements were so inconsistent that no longitudes could be determined.

And so Velasco changed his strategy. Instead of relying on hundreds of local people scattered across the countryside to make measurements of a lunar eclipse, he would send an experienced astronomer to Mexico City to ensure that at least one valuable measurement would be made. He chose Jaime Juan who, in writing the justification for the selection, Velasco described as "an expert in mathematics and astronomical calculations." Almost nothing more is known about Juan except that he was working in Valencia at the time of his selection and he was put in debtors' prison in the summer of 1583. He would have missed sailing to the New World if the departure of the Spanish fleet that year had not been delayed by the construction of new galleons.

But sail he did, and he carried with him a detailed set of instructions prepared by Velasco and his advisers that told him how to construct an instrument to observe the eclipse by using "the straight side of a plank that is at least a yard wide and long" and "a thin iron rod, one-third yard long" with "a thin thread with a loose knot tied to the rod" and so forth. The instructions also specified the wording Juan was to use when he reported his observations, to ensure maximum accuracy and clarity. He was also to do a variety of other types of measurements wherever he made landfall. He was to spend several days observing the path of the sun to determine the direction of true north and mark it on a large stone slab. Then he was to compare that direction to the orientation to magnetic north indicated by a compass to determine the magnetic variation that would be useful in navigation. He was also to instruct local captains on new navigational techniques and demonstrate six new navigational instruments that Velasco had given him. Given the disdain Spanish sea captains had for any type of navigation except dead reckoning, this last task proved to be the most challenging.

When he arrived in Mexico City, he became acquainted with three local men who would assist him in observing the eclipse. One was Francisco Dominguez de Ocampo, a cartographer who had been sent to New Spain years earlier to make detailed maps. Pedro Farfán was a local doctor and naturalist who was collecting animal and plant specimens, describing them, and then sending them to Spain for others to examine. The third and, probably, the most valuable man was Cristóbal Gudiel. He was a gunsmith and the first person authorized by the government in Spain to produce gunpowder in the New World. His skills were invaluable in that he was given the delicate task of transporting and setting the weights on a mechanical clock loaned by the local archbishop so that Juan could time the lunar eclipse.

The four men met for several days before the eclipse to practice their routine. It was decided that each man would make an independent estimate of when the eclipse began and when it ended. On the afternoon of the eclipse, November 17, 1584, according to Juan, the clock was "set as well as could be done." Then the four men waited.

According to the prediction provided by Velasco, the eclipse should start about an hour after moonrise. Instead, the Moon rose already darkened. And so measurements could only be made at the end of the eclipse.

The Moon was still close to the horizon when it began to leave the Earth's shadow, further complicating the ability to determine exactly when the eclipse ended. Nevertheless, the four men made their determinations. Juan wrote his report and sent it to Spain where Velasco had his advisers do the necessary calculations. The result was not as good as hoped—a combination of an unreliable clock and a poor eclipse prediction—but it was an improvement.

All previous attempts to locate Mexico City and determine its longitude gave uncertainties of almost 15°, that is, more than a thousand miles. Juan and his collaborators managed to reduce that to about two hundred miles. It improved the general accuracy of maps, but the location of the coastline of New Spain was still unknown by that amount.

After Mexico City, Juan traveled to the western coast where he joined a ship of the Spanish fleet that would be sailing across the Pacific. The fleet first went north exploring the California coast as far north as Mendocino. Then it turned west and headed for Manila in the Philippines where, following Velasco's instructions, he was to observe a lunar eclipse set to occur in 1587. Yet soon after arriving in Manila, he became sick and died of a fever before observing the eclipse, and thus Velasco's efforts were foiled once again.

—•—

As already hinted, the lessons learned from Jaime Juan and his efforts at Mexico City were twofold. First, in order to use a lunar eclipse to make an accurate determination of longitude, one had to have a reliable clock. And the first such clock would not be invented until well into the next century when Dutch mathematician and horologist Christiaan Huygens designed and built the world's first pendulum clock. Second, the prediction of lunar eclipses had to be improved. The ones based on Regiomontanus's *Ephemerides* or Zacuto's *Almanach Perpetuum* or the dozens of similar tables—there were now many such tables because of the proliferation of printing presses—could be in error by hours. And so there was a reluctance to send someone to a distant land to observe an eclipse when, in the end, it might not be visible or be missed. In addition, because a clock had to be set by making astronomical observations, improved tables meant the resetting could be done more reliably.

This was also a time when Ptolemy's view of the universe was being challenged. In the first book of the *Almagest*, he described a geocentric system whereby the Earth is at the center and the Sun, the Moon, and the planets orbit around it in circular orbits. He included a secondary motion for each body by adding an epicycle to each orbit, that is, the motion of each heavenly body was the combination of two circles, a large one centered on the Earth and a smaller circle, or epicycle, with a center that lies along the path of the first. This was to account for the slight variation in rate that was observed when one of these bodies was followed against the background of stars.

In 1543 Nicolaus Copernicus published *De revolutionibus orbium coelestium* ("On the Revolutions of the Celestial Spheres") in which he proposed a heliocentric system. According to Copernicus, the Sun was at the center, and the planets, including the Earth, revolved around it in circular orbits, each one with an epicycle. He also proposed that the Moon was in orbit around the Earth and that it followed a circular path also modified with an epicycle.

In 1551 Erasmus Reinhold published the first astronomical tables based on Copernicus's heliocentric system. Because the Duke of Prussia financed Reinhold's work, the published tables are known as the *Prutenic* (or Prussian) *Tables*. Given that it used a heliocentric system instead of a geocentric one, one might expect the *Prutenic Tables* to be an improvement over all previous work. It was not. The ability to predict planetary positions—or eclipses—remained unchanged because Copernicus was still proposing that the planets, including the Earth, moved around the Sun in circular orbits. It would be another seventy-five years before a major improvement was made. And it would be made by an astronomer who, considering the nature of the profession, ironically suffered by poor vision.

Johannes Kepler would always remember the two events from his boyhood that pointed him toward his later calling. The first occurred when he was five years old. It was a winter evening and his mother led him up a slope and showed him a great comet then in the sky. The next occurred three years later when his father took him out at night to see a lunar eclipse.

Kepler was educated at a university at Tübingen in southern Germany. His professor of astronomy was Michael Maestlin, one of the first people to accept and teach the heliocentric Copernican view. In 1600 Kepler began working as an assistant to Tycho Brahe, who was the last great naked-eye astronomer, making his observations of the night sky just decades before the invention of the telescope. Brahe, however, was a firm believer in a geocentric universe. During the many years that he studied the sky, he had never seen the stars shift back and forth, which they might do if the Earth made an annual orbit around the Sun. Also there was no indication that the Earth was going at a fantastic speed, something that was also thought to be a requirement if Copernicus was correct. And, most importantly, a universe with the Earth at the center was consistent with biblical scripture; otherwise, so it was argued, during the Battle of Gibeon, Joshua would not have had to stop the movement of the Sun because it would have been at the center and stationary.

Brahe actually proposed a modification of Ptolemy's geocentric system. Brahe's system also had the Earth at the center, but the Sun and the Moon revolved around the Earth and the five visible planets around the Sun. And Brahe intended to compute a set of astronomical tables for his system. He hired Kepler to do the calculations, but just as the work was beginning Brahe died. On his deathbed, Brahe begged his assistant to complete the work by following his system, not the Copernican one.

Kepler spent the next years in torment. He was a man who was always ill. He suffered from multiple vision in one eye. Boils and other violent eruptions appeared on his body, especially the shoulders. He found it difficult to sit for a long time; he had to keep moving back and forth. He resorted to gnawing bones and eating dry bread. For years, he tried to get the computations for Brahe's system to match the observations that Brahe had made over the decades. But it was impossible. He changed to the Copernican system, and still there were wide discrepancies, especially in the motion of Mars. In December 1604 he was weighed down by thoughts of death and considered abandoning his work and letting some future mathematician take up the challenge. But, then, the next Easter, he instantly saw the solution.

The real and the computed motions of Mars matched if the planet's orbit was an ellipse, not a perfect circle. And what was true for Mars, he realized, must also hold for the other planets. That meant all the planets, including the Earth, followed elliptical paths around the Sun with the Sun at a focus. It was, Kepler would say, as if he had been awakened out of a deep sleep.

He delved further into the calculations, discovering that the speed a planet moved along its elliptical path depended on its distance from the Sun. The farther away a planet was, the slower it moved. He was able to quantify it: The speed of a planet varied in such a way that a line drawn from the Sun to the planet swept out equal areas in equal times.

In 1609 he published the discovery of elliptical orbits and planetary speeds in the book *Astronomia Nova* ("The New Astronomy"). Then he began the task of computing a new set of astronomical tables. It took him eighteen years. The following year he claimed the money promised to him by the Emperor of the Holy Roman Empire, Rudolf II, to print the tables. Once the money was acquired, he went to the city of Linz to have the tables printed, but the city was under siege as part of the Thirty Years' War and his manuscript was almost lost. Through all this Kepler received letters as far away as India and China from Jesuit astronomers who had heard of his work and wanted a copy of the tables. Finally, he found a printer in the city of Ulm, Jonas Saur, who Kepler considered to be "unpleasant, proud, extravagant and impetuous." Nevertheless, he negotiated terms with Saur and had the tables printed, naming them the *Rudolphine Tables*, after the Holy Roman Emperor.*

---

* It should be mentioned that during these years, Kepler also married and intended to have the wedding on the day of a total lunar eclipse, October 18, 1613. For a reason that was never given, days before the planned event, the wedding was postponed and took place two days after the eclipse.

As already noted, there was much interest in Kepler and his work as he was doing his calculations for the *Rudolphine Tables*. Among those interested was Henry Gellibrand who, sometime after being appointed professor of astronomy at Gresham College in London, acquired a copy of Kepler's *Astronomia Nova*. He read it intently, leaving annotation marks on almost every page. And so, when the *Rudolphine Tables* was published in 1627, it is likely that Gellibrand acquired a copy and studied this book too. That may have led him to realize that a new effort should be made to use a lunar eclipse to measure a distance across the globe, using instead as a foundation Kepler's theory that the Earth orbited the sun in an ellipse, not a circle. But he needed to find someone who would be making a trip.

Little is known about the early life of Captain Thomas James except that he was Welsh and that his father had been an alderman and twice mayor of Bristol, England. He first entered the public record in 1628 when letters of marque were granted to him to act as a privateer and captain of the *Dragon*, a 140-ton ship, six years old, with ten guns. The purpose was to guard the British coast from French frigates. His next appearance in the public record was in 1631 when he was chosen by members of the Society of Merchant Venturers in Bristol to lead a voyage to search for the Northwest Passage.

Ever since the 1492 voyage of Columbus, Europeans had been trying to bypass the continents of America to get to India and China. English explorer Henry Hudson made a famous attempt in 1611 when he sailed as far north as he could, limited by a wall of ice, then to the west where he discovered the large bay that now bears his name. Twenty years later, Thomas James was going to repeat the effort.

James decided that sailing with one small ship of not more than seventy tons would be best. He chose the *Henrietta Maria*, named for the British queen. He decided also to limit the crew to twenty-two men who must be "unmarried and healthy," and refused all who applied who "had used the northerly icy seas" lest one of them question a decision and undermine his authority. He loaded the *Henrietta Maria* with provisions for eighteen months. He also acquired for the voyage "a Chest full of the best and choicest Mathematicall bookes that could be got for money in England," and bought the best navigational instruments and sought out those who were known to give the best advice on how the instruments should be used. It was seeking such advice that led James to Professor Gellibrand at Gresham College.

In their conversation, Gellibrand undoubtedly told the captain of the remarkable opportunity they would have if the two men made simultaneous observations of a lunar eclipse, Gresham near London and James in northern North America. But there was a problem. The next such eclipse would occur in November, and so James would have to winter in the arctic. He agreed to do it.

The *Henrietta Maria* sailed from Bristol on May 13, 1631. Captain James sighted Greenland on June 14. The ship was surrounded by ice floes the next day. By June 27 the entrance to Hudson Bay was reached.

Sailing through an ice-filled sea, James reached the southern extreme of Hudson Bay on September 12. For a month he sailed close to the coastline, hoping to find a large river that might lead close to the St. Lawrence River, but none was found. By mid-October, the onset of winter unmistakable, he decided to winter at Charlton Island. The crew built a storehouse, a cookhouse, and three huts for sleeping, using sails to form the roofs. James spent two weeks making measurements of the Sun's position at noontime, determining, correctly, that his position was only thirty miles north of London's latitude.

According to his log, on November 8, 1831, he "observed an eclipse of the Moon with what Care possible I could, both in the Trial of the Exactness of our Instruments, as also in the Observation." He also recorded "the Land was all deeply covered with snow."

He was now weather-bound, and so, a month later, with the weather continuing to worsen, he decided to sink the *Henrietta*

Maria to keep it from being crushed by the rising mountains of ice. He and the carpenter went down in the hold, took an auger and bored "a hole in the Ship, and let in the water." James and his crew now wondered whether they would survive the winter.

But they did. By May the ice had noticeably receded. With iron bars, knives, and hot water the crew began to dig out the sunken *Henrietta Maria*. Two months of hard labor freed the ship. The carpenter fixed the holes. And all prepared to leave. But first a ceremony was held on the beach next to a large cross they had erected to mark the graves of four comrades who had died that winter. On July 11, 1632, the *Henrietta Maria* sailed. On November 1, the ship was back in Bristol.

James wrote an account of his ordeal, *The Strange and Dangerous Voyage of Captaine Thomas James,* published in 1633. It was the first best-selling narrative of a polar expedition and was later to influence Samuel Taylor Coleridge's *The Rime of the Ancient Mariner.* Attached to the account was an appendix by Gellibrand who reported on the measurements of the lunar eclipse.

Gellibrand concluded that the difference in longitude between Gresham College in London where he made his measurements and James's campsite on Charlton Island was 79° 30'. The actual difference, according to today's maps, is 79° 45'. To put this in terms of mileage: the actual distance between Gresham College and Charlton Island is 3,218 miles. Gellibrand and James had determined a distance of 3207 miles, a difference of only *11 miles!*

It was an astonishing achievement. Gellibrand and James knew they had accomplished something significant, Gellibrand estimating that the uncertainty in the distance he calculated was probably about twenty miles. Signaling the achievement, at the end of the appendix, Gellibrand wrote that this should "be means to encourage our English Sea-men and others to make such or the like observation in foreign ports, as the heavens shall be offered unto them."

———•———

Kepler's *Rudolphine Tables* continued to show their superiority in making predictions over previous tables. It was used to give the first prediction of a transit of the planet Mercury across the Sun, which was observed on November 7, 1631, by Pierre Gassendi in France. It also gave the first prediction of a transit of Venus, which Jeremiah Horrocks observed from England on December 4, 1639. It allowed Nicolas-Claude Fabri de Peiresc of Paris to organize observers in eight different cities spread across the Middle East, the Mediterranean, and Europe to time the lunar eclipse of August 28, 1635, and, from those timings, correct Ptolemy's estimate of the length of the Mediterranean Sea. And there was much more.

The *Rudolphine Tables* were the first set of astronomical tables that allowed one to calculate the future positions of planets and the Moon with an accuracy that permitted accurate predictions. People soon understood that this could also be done in the reverse. That is, the *Rudolphine Tables* could also be used to predict *past* positions of the planets and the Moon. And that opened the door to the possibility of determining the exact dates of historical events, that is, when a historical event could be connected to a celestial event, such as an eclipse.

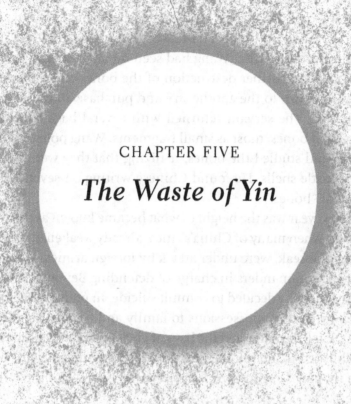

CHAPTER FIVE

The Waste of Yin

The eclipse prediction is wrong again. If there is a mistake next time, you will be punished severely.

—a message from Emperor Chongzhen to the
Chinese Bureau of Astronomy, 1629

In 1899 the people of Beijing, China, were besieged by an outbreak of malaria. One of those afflicted was Wang Yirong, the director of the Imperial University. He consulted a physician who prescribed a well-known remedy to curb the high fever. Wang was to send a servant to an apothecary to procure a bag of dragon bones. The bones were then to be ground up into a powder, added to a hot tea to a make a poultice, and then applied to the most affected areas of the body.

Wang followed the advice. He did send a servant to obtain a bag of dragon bones. Then, as the bones were being ground into a powder, Wang noticed that a few of them had Chinese writing on them similar to what Wang had seen on ancient bronze vases. He stopped the further destruction of the bones and ordered the servant to return to the apothecary and purchase all of the bones that he had. The servant returned with several bags containing hundreds of bones, most as small fragments. Wang poured out the contents and studied the bones, realizing that they were actually ancient turtle shells. He found Chinese writing on several of the additional "bones."

The next year was the height of what became known as the Boxer Rebellion when many of China's cities, already weakened from the malaria outbreak, were under attack by foreign armies. Wang was one of the commanders in charge of defending Beijing. When the defense failed, he decided to commit suicide. In preparation for his suicide, he gave his possessions to family and friends. To another scholar, Liu E, he entrusted the bags of ancient turtle shells. In 1903 Liu published in a book the first description of the ancient shells, which by now were known as *oracle bones* because of the writing on them.

Interest in these and other oracle bones grew. One person in particular, Luo Zhenyu, an expert on ancient Chinese culture, amassed one of the largest collections. He also made the first major breakthrough in understanding where and when the writings on the oracle bones had been done. He recognized in the writing several names of kings of the Shang Dynasty. And he knew where the capital of the Shang Dynasty had been located: about three hundred miles south of Beijing along the Yellow River in a region known as Yinxu, "the Waste of Yin."*

China was still in political turmoil as the effects of the Boxer Rebellion reverberated across the land. And Luo, a supporter of one of the revolutionary armies, had to keep moving. It was not

---

* Here the word "waste" refers to the many ancient ruins in the area.

until 1915 that he managed to make his first trip to Yinxu to see if he might be able to find oracle bones that where still buried in the ground. Fortunately, he kept a diary of the trip so that we know where he went, what he saw, and what he was thinking.

It was May 13, 1915. He arrived by cart and immediately began to question local villagers about the place where shells might be found. They showed him an area of about forty acres. He walked over it, picking up numerous shells, also some bones, many covered with writing. He also found bone arrowheads, ivory daggers and hairpins, stone knives and axes and bone writing tablets. He found animal remains, in particular, elephant tusks and teeth, things that are rarely found in China today. But what interested him most were the so-called oracle bones. What was the purpose of the writing? And what did it say?

———•———

Archaeological excavation of Yinxu began in 1928. The work continues today. So far, more than twelve square miles have been excavated, revealing building foundations, numerous burial and sacrificial sites that contain animal and human remains, and of particular concern here, pits filled with oracle bones. To date about 200,000 fragments have been uncovered, of which some 50,000 bear inscriptions.

The site does date from the Shang Dynasty, that is, from about the fifteenth to the thirteenth centuries B.C.E. Which means the writings inscribed on the oracle bones are over three thousand years old, making them among the oldest—perhaps the very oldest—examples of extensive written material known to us. It is also written in the Chinese language, which by good fortune has not changed appreciably since then, and so, with a high measure of certainty, it can be translated.*

---

* About six thousand different characters have been recorded, of which some two thousand can be identified with modern versions. Most of the remainder are probably proper names.

The oracle bones themselves are usually pieces of turtle shells or animal bones, most often the scapulae of ox. The writings record questions to which answers were sought by divination. During such a divination ceremony, after an appropriate question had been inscribed, a shell or bone was subjected to the intense heat until it cracked. Then the pattern of cracking was interpreted. After the ceremony, the shell or bone was placed—perhaps thrown—into a storage pit.

The questions cover a wide range of subjects. Some are questions about whether to perform religious rituals or sacrifices or ask the meaning of dreams. Others relate to future warfare or hunting. Still others are concerned with journeys that are to be made by kings or about the birth of children. Some even record the health of royal people and their ailments. Natural events are also recorded, such as the occurrence of heavy snow or a strong wind. And then there are those oracle bones that record the occurrence of celestial events, such as eclipses.

Of particular note is an inscription that has been translated as: "Three flames ate the Sun, a big star was seen." Many have interpreted this as a record of a total solar eclipse. The "three flames" would be solar prominences, that is, large gas jets in the solar atmosphere that can only be seen during an eclipse, and the big star is either a planet or a bright star that stood out once the Sun was momentarily darkened. Calculating backward the motions of the Sun, the Moon, and the planets to when a total solar eclipse was seen from Yinxu with a planet or bright star near to the Sun—also constraining the event to occur during the Shang Dynasty—gives the date June 5, 1302 B.C.E. In this case, the bright "star" would have been the planet Mercury.

Thirteen eclipses, six solar and seven lunar, have been identified in inscriptions on oracle bones. A few of these have proven useful in establishing exact key moments in Chinese history.

Toward that end, in 1996, the Chinese government established the Xia-Shang-Zhou Chronology Project to build a credible chronology of early China during the Xia, Shang, and Zhou dynasties.

The project employs more than two hundred specialists in fields that range from history and linguistics to archaeology and astronomy. Naturally, the dating of celestial events mentioned on oracle bones is part of the project.

There are complete lists of kings for the Xia, Shang, and Zhou dynasties. But only relative dates were known of when each king ruled. To determine the exact years, at least one absolute date had to be established in the lists of kings. The one that was the most promising was the date when King Wu Wang of the Zhou success-fully led an army of war chariots and thousands of troops against the army of Shang King Di Xin, defeating Di Xin and marking the end of the Shang Dynasty.

According to classical writings, the defeat occurred sometime between 1130 and 1018 B.C.E., a range of 112 years. Workers of the Chronology Project acquired other evidence, which included the dating of Zhou pottery shards using carbon-14 techniques, to reduce this to a thirty-year period between 1050 and 1020 B.C.E. Astronomical observations were then used, including some found on oracle bones, to determine an exact date of the defeat of the Shang king: January 20, 1046 B.C.E. That date is the beginning of the Zhou Dynasty, a major event in Chinese history, which now allows one to determine an absolute chronology for the list of kings.

This is all to illustrate how historical chronology is used today. And how it still very much relies on ancient astronomical records. The procedure followed recently by the Chronology Project in China is exactly the one outlined by those who began this unique field of study at the time of Kepler, nearly three centuries ago.

———•———

How do we know that Julius Caesar was born on July 13, 100 B.C.E. and that he lived fifty-six years and died on March 15, 44 B.C.E.? Or that Confucius died in 479 B.C.E. or that Buddha died four years earlier in 483 B.C.E.? Or that the first ancient Olympic games were played in 776 B.C.E.? It is by combining chronological references found in ancient sources with the precision of astronomical calculations to determine absolute dates. And eclipses are the cornerstones of these calculations because they are spectacular events and they are often recorded.

While the method is straightforward in practice, it is painstakingly tedious. In antiquity, there was no single, universally accepted calendar system. In Greece alone, evidence has been uncovered for more than a thousand different calendar systems.* (It seems that not only every city-state had its own civil calendar, but that every temple had its own calendar to determine when specific rites should be performed or festivals should be celebrated.) And, among these different calendars, there was no agreement as to the beginning of the year or when a month began. To complicate it further, civil calendars in Greece did not follow any set of rigid rules, as suggested by the complaint of the gods in Aristophanes's *The Clouds*, the Athenians failing to feed them in a timely way.

But there is a saving grace. The Greeks and the Romans, as well as the Chinese, were obsessed with maintaining lists of rulers and recording the lengths of their reigns. And these same civilizations also kept eclipse records. Furthermore—and this has been one of the greatest benefits—the early Christian writer Eusebius compiled in parallel columns a list of the rulers of Athens and Rome and the earlier rulers of Assyria and Israel that begins in the time of Abraham and continues to Eusebius's time in the fourth century C.E.

Examining sources from around the world, for the time interval from 700 B.C.E. and 1600 C.E., there are about three hundred eclipses with sufficient information to determine accurate eclipse times. A few are easy to determine. For example, from a report in England dated to the eighth century C.E., the Moon moved in front of Jupiter at the time of a lunar eclipse. Because this type of event is unusual—it happens only once every several hundred years— this particular one must have occurred on November 23, 755 C.E. In China, as recorded in an ancient source known as the *Bamboo*

---

* There are about forty different calendars in use in the world today.

Annals, it is stated: "In the first month of spring, in the year King Yi ascended the throne, the day dawned twice at Zheng." The "day dawned twice" is probably a solar eclipse that occurred soon after sunrise. For a given spot on the Earth, such an event is rare. And so this can be matched, with great confidence, with a solar eclipse that occurred after sunrise as seen from "at Zheng" (modern Weinan in the Shaanxi Province in central China), on April 21, 899 B.C.E.— which must also be the date when King Yi, the seventh king of the Zhou Dynasty, ascended to the throne.

Kepler was the first to use his new ideas about elliptical orbits of the planets and the Moon to determine the dates of ancient events. In fact, he made a determination of the year of Christ's birth while he was computing the *Rudolphine Tables*.

More than a thousand years before Kepler, at the request of Pope John I, the monk Dionysius Exiguus, living in Rome, prepared a table of the future dates of Easter. At the time it was customary to use the accession of the Roman Emperor Diocletian in what today is identified as 284 C.E. as the basis to begin a counting of years. Rather than "perpetuate the name of the Great Persecutor," Dionysius decided he would "number the years from the Incarnation of our Lord Jesus Christ." From his knowledge of ancient history, he reckoned the birth of Jesus had occurred in the year 753 from the founding of the city of Rome. And that is how it stood for three hundred years until a monk in Germany, Regino of Prüm, was writing a chronology of Biblical figures and discovered that Christ must have been born years earlier than Dionysius had determined. But how much earlier? That was the question Kepler had decided he would answer.

He knew, from a passage in the book of Matthew, that Christ was born during the reign of King Herod. He also knew that, according to the Jewish scholar Flavius Josephus, who wrote in the first century C.E., Herod died soon after a lunar eclipse and before Passover. Kepler did the calculations. He determined that the lunar eclipse that preceded Herod's death must have been the one that occurred on March 13, 4 B.C.E. Based on other evidence of Christ's life, such as knowing where he preached, Kepler determined that Christ must have been born in either 4 or 5 B.C.E. It is a conclusion that many have challenged, but no one has improved on. Today most experts agree that Kepler's determination is the most plausible.

Commonsense rules to determine an ancient date were first stated by one of Kepler's contemporaries, the French Jesuit scholar Dionysius Petavius. Petavius was well suited to provide such guidance. He was skilled at what was regarded as the learned languages of Greek and Latin and was deeply versed in history. And he was a good mathematician and astronomer. According to the well-skilled and well-read scholar, the determination of an ancient date required three things. One must begin with a reliable source or authority. Next the chronologist had to carry out the exacting and quite laborious task of calculating a set of eclipses. Finally, after comparing the set of eclipses with information from the authority, one must be able to narrow the set down to a single event, then compare the timing of that event to other, less reliable and more ambiguous sources to see if everything was consistent. Petavius showed how one should proceed by demonstrating how he had determined the starting year of a famous event in ancient history, the Peloponnesian War.

For his authority, Petavius chose the Greek general and historian Thucydides. Thucydides had been in the war on the side of Athens against Sparta. And he had written a detailed account of the conflict. Within the account, as Petavius knew, Thucydides had mentioned three eclipses.

The first one was a solar eclipse that the Greek general said had occurred after midday in the summer a few months before the start of the war. He also mentioned that stars became visible during the eclipse, which meant it was either total or nearly total. The only total solar eclipse that passed close to Greece during that period and that occurred after midday and in the summer was on August 3, 431 B.C.E. Furthermore, the path of totality passed just north of Thrace in northeastern Greece. Thucydides had an estate in Thrace and had the right to work a gold mine in the area, and so he probably viewed the eclipse.

His comments about the second solar eclipse are terse. He merely says that it happened during the eighth year of the war. That would be in either 423 or 424 B.C.E. And there is a partial solar eclipse visible from Greece on March 21, 424 B.C.E.

The lunar eclipse is by far the most interesting of the trio of eclipses because it is associated with a fateful decision made during the war.

Athenian forces had laid siege to the Spartan colony of Syracuse on Sicily for more than two years. During that time, little progress had been made, and so the Athenian commander, Nicias, decided to return to Greece and defend the mainland from further Spartan expansion. On the night before the planned departure, there was a lunar eclipse, and Nicias asked the priests what he should do. They suggested the Athenians wait thrice nine days, that is, twenty-seven days after the eclipse. Nicias agreed. The defenders of Syracuse went on the attack and killed 40,000 Athenians. As Thucydides later wrote: "This was the greatest Hellenic achievement of any in this war, or, in my opinion, in Hellenic history; at once most glorious to the victors, and most calamitous to the conquered." Athens was eventually defeated by Sparta. From the information provided by Thucydides, Petavius dated the lunar eclipse to the night of August 27, 413 B.C.E.

Petavius then showed how he could use other statements by Thucydides to bootstrap his way to determine the dates of other key events in Greek history. For example, Thucydides mentioned the eminence of an early Greek named Cylon who was a winner of an Olympic games. Then, with a list of winners of past Olympic games compiled by the third century B.C.E. Greek philosopher Eratosthenes, Petavius could determine that the first games must have been played in 776 B.C.E. And he found support for this date in the writings of other Greeks.

Though Petavius and later chronologists report their findings in a linear manner pegged to a timeline, what they have actually produced is a web that connects dates and facts. Shift one node of the web, that is, change one date, even slightly, and a whole section of the web could change. But that is the intrigue of historical chronology. A small army of modern chronologists are at this moment sifting through archives and wondering if the next scroll of papyrus they unroll or the next sheet of medieval vellum they study might contain a key element that will make the web sturdier or cause it to shift by a large amount.

And if you asked any of these current researchers which chronological event was most important in anchoring the present timeline of ancient history, you would receive a unanimous response. It was the Nineveh eclipse.

Of the thousands of clay tablets taken from the ancient Assyrian cities of Nineveh, Assur, and Sultantepe in modern-day Turkey and Iraq, nineteen tablets stand out. On those nineteen is a list of officials whose names are attached to their titles and to the principal events that took place during their rule. Where one tablet is broken away, one or more other tablets provide the necessary overlap. In all, there are 261 names. But when did these people serve?

Considerable work showed how the list of names on the nineteen tables, known today as the Assyrian Eponym Canon, could be matched with lists of names of people who ruled concurrently in Judah and Israel. But the same problem remained. When did these people serve? How much time was there between the first name on the Canon and the last? And that is where the matter stood until a portion of the cuneiform text—apparently important because the entry had a line drawn across the entire tablet—was transliterated as *shamash antalu*. The first word, shamash, was quickly deciphered as "the Sun." But what of antalu? As is common in language, this word had multiple meanings. It could mean to bend or to twist or to contort. It also meant to cover or to make obscure. That was the key. Shamash antalu was a direct reference to a solar eclipse.

As already mentioned, the list of names—and events—on the Canon could be correlated with names and events in other ancient sources, including the Bible. In fact, there may be as many as eight biblical references to this particular solar eclipse in the books of the Hebrew prophets. In Isaiah 13:10, it is recorded: "The rising sun will be darkened and the moon will not give its light." And in Amos 8:9: "On that day, says the Lord, I will make the Sun go down at noon and darken the earth in broad daylight." The second quotation is especially revealing because the book of Amos begins: "These are the words of Amos . . . two years before the earthquake." And there are other passages that relate this eclipse and an earthquake, for example, in Joel 2:10: "Before the earth shakes, the heavens tremble, the sun and moon are darkened, and the stars no longer shine."

Archaeological work supports a major earthquake striking near the Sea of Galilee during the time of the Hebrew prophets. King Solomon's palace in Jerusalem was damaged and repaired. Walls collapsed in the ancient cities of Hazor and Samaria. One can imagine that this earthquake coming just two years after a solar eclipse left a strong impression on the people of the Middle East and is the reason both are recorded. No less than five prophets used this coincidence to threaten sinners with the divine wrath. But, again, we want to know when these two events occurred.

The answer, once we turn toward eclipses and astronomical calculations, is easy to determine. Over a span of several hundred years, only one eclipse matches the one recorded on the Assyrian Eponym Canon: a total solar eclipse that passed over what is now Israel, Syria, and northern Iraq on June 15, 763 B.C.E. This is the anchor point that assigns an absolute chronology to the lists of kings for Assyria, Judah, and ancient Israel. We now know the lists of kings for Assyria runs from 911 to 612 B.C.E. And that list has now been intermeshed with lists of Macedonian, Greek, and Roman rulers to provide an absolute chronology for the ancient Western world.

What is amazing is that, through this effort—chronologists have been at their jobs for the last few hundred years—one can now open a book and read of an ancient event and be confident of the date assigned. We are no longer limited simply to wandering through ancient ruins and wondering about their ages, wondering when exactly Alexander the Great walked through these streets. We can now see these same ruins and understand these ancient persons through the additional perspective of absolute time.

If one now returns to the seventeenth century and the time of Kepler, one might think that the birth of modern chronology represented the major influence that eclipses had on this period of history. But that would be wrong. Remarkably, a sequence of eclipses influenced world politics. To understand that story, one has to step back once more to China.

——•——

The oldest written record of an observation of the night sky is from China. It predates the oldest inscription on an oracle bone by hundreds of years. It is contained in a collection known simply as the *Book of Documents* and records how star patterns could be used to determine the seasons.

As in many civilizations, the Chinese were interested in watching the sky because they believed the movements of stars and planets were closely related to the destiny of the country and its rulers. And to follow these movements, they constructed calendars.

The earliest Chinese calendar dates from the fourteenth century B.C.E. and was maintained by the Yin people of the Shang Dynasty, the same people who may have made some of the oracle bones that fascinated people in the wake of the Boxer Rebellion and that centuries later would help modern Chinese scholars reconstruct history. Their calendar consisted of twelve moons, or lunar months, of either twenty-nine or thirty days each with an additional month added every few years to keep the calendar synchronized with the seasons. (We do the same thing by adding a leap day every four years to our modern calendar.) In total, the Chinese constructed more than one hundred different calendars. And each calendar had many minor adjustments. A main purpose was to predict when a celestial anomaly, such as an eclipse, would happen in the future so that a specific ceremony could be prepared in advance. In that way, the rulers of the current dynasty could demonstrate as spectacularly as possible that they possessed a heavenly mandate.

The calendar in use in China when Kepler was doing his work, that is, during the Ming Dynasty, had been in use for nearly three hundred years. During those three hundred years, minor complications in the motion of the Sun and the Moon had grown to such a degree that those who maintained the calendar were uncertain when to insert an additional month.* By the end of the sixteenth century, a crisis had arisen. Predictions were missing eclipses by hours. The straw that broke the camel's back came in 1592 when the Ministry of Rites, which was charged with making the predictions, erred by a full day in predicting an eclipse. This was not only an unfortunate mistake, but also a grievous one. Runners had not been sent in time to inform the local population. And the emperor had not performed the proper ritual before the eclipse. By coincidence, just a few years earlier, the first Jesuit priests had arrived from Europe and they maintained that they knew how to improve the Chinese calendar.

The Society of Jesus, its members known as Jesuits, was founded in 1534 by Ignatius of Loyola and received papal approval in 1540. Members of course were required to follow strict rules of discipline and to take a special vow of obedience to the Pope. They were also required to cultivate scholarship among themselves and to stress learning. Quite a few became eminent mathematicians and astronomers. Those who entered China in the late sixteenth century brought with them many books that, after they learned the language, they translated into Chinese, including Euclid's *Elements of Geometry*. They were willing minions of Chinese rulers because, as their letters back to Europe indicated, they considered this approach

---

* A major cause of the uncertainty was precession, a slow cyclic movement of the Earth's spin axis as it is projected on the distant background of stars. Precession is caused by the Sun's gravitational attraction on the Earth's equatorial bulge that makes the Earth's axis move much like a spinning top does when placed on the ground at an angle.

to be the most effective one to sanction their religion in China and gain inroads to converting the population that way.

Their acceptance in China was slow. The early seventeenth century was a period of political decline. Battles between government and bandit armies were common. The ruling emperor, Wanli, was inept. Extortion was common among provincial officials. And problems with the calendar continued.

Finally, in 1610, the Jesuits in Beijing were asked to make a prediction of the timing of the beginning of a partial solar eclipse that would occur in December. The official Bureau of Astronomy and Calendars within the Ministry of Rites also made one. The Bureau's prediction, though good by the standards of the day, was off by thirty minutes. The Jesuits' prediction, based on recent observations of the Moon and the use of spherical trigonometry—neither were used in the Chinese prediction—was closer. A second comparison of predictions was made in 1612 during a partial lunar eclipse, and again the prediction made by the Jesuits was closer.

In 1629, the year after a new emperor, Chongzhen, assumed power, a third comparison of predictions was made. This time the Jesuits relied on calculations made by Johann Schreck, a Jesuit who had arrived in China in 1623 from Germany and who knew Kepler personally.

Soon after his arrival, seeing the situation in China and the importance of eclipse predictions, Schreck wrote to Kepler asking for advice. Kepler's response arrived in 1627. In it, he outlined how an eclipse prediction could be improved by using an elliptical orbit for the Moon's orbit. He also included a copy of the recently printed *Rudolphine Tables*.

The test came on June 21, 1629, during a partial solar eclipse. Schreck had predicted the silhouette of the Moon would first touch the bright disk of the Sun an hour earlier than the official Chinese prediction. Chongzhen sent an admonishment to the Bureau of Astronomy and Calendars. "The solar eclipse prediction by the Bureau is inaccurate again," his message said. "If there is a mistake next time, you will be punished severely." To insure that a future mistake was unlikely, he informed members of the Bureau that he was appointing Schreck and other Jesuits to work with them.

Schreck died the next year. The Jesuits replaced him on the Bureau with Johann Adam Schall, a Flemish Jesuit who had assisted Schreck in eclipse calculations. Schall was assigned a team of more than fifty Chinese mathematicians and astronomers to work with him at the Bureau. Over the next ten years, they compiled a staggering forty-six titles in astronomy in a total of 137 volumes, producing a massive compendium entitled *Books on Calendrical Astronomy of the Chongzhen Reign*. This was the formal introduction of Earth-centered, Western, or Copernican, astronomy in China. Then Schall and his Chinese colleagues set upon the task of producing a new calendar.

The new calendar was ready by early 1644, but by then rebellion was in the air. For centuries the Great Wall, built during the early years of the Ming Dynasty, had kept northern invaders from conquering China. And the wall continued to serve its intended purpose. The defeat of the Ming Dynasty did not come from without, but from within.

After a year of bloody fighting, the rebel leader—or bandit, depending on one's view of history—Li Zicheng took over Beijing on April 24 with little resistance. The next day Chongzhen hanged himself, ending the Ming Dynasty.

On May 25, the Manchu leader Wu Sangui and his army crossed through the Great Wall at Shanhai Pass, and twelve days later took control of Beijing, executing Li Zicheng. Thus began the Qing Dynasty in China. The Manchu emperor, Shunzhi, was now also the emperor of China. He was seven years old.

Because of Shunzhi's youth, a council of Manchu warriors now ruled China. The council ordered all residents and businesses located near the imperial palaces contained within the Forbidden City to vacate and move south so that Manchu soldiers and their families could take up residence around the court, creating a vast armed garrison. Many of the Jesuit churches were within the vacated area. Schall sent a message to the council asking that he

be permitted to reside and maintain his church within the area.
He also told them of the new calendar he had prepared, writing:

> The imperial calendar of the Ming Dynasty was terribly
> wrong in 1629 and I was ordered to improve it. I compiled
> more than 140 chapters on mathematical astronomy that
> can be used to predict solar and lunar eclipses. It was my
> great fortune that the good Qing Dynasty expelled rebels
> from the land. Thanks to the greater glory of God the
> Qing Dynasty came to our rescue.

The Manchu leaders took notice of Schall and his request and of
the possibility of a new calendar, but they were wary of all Europeans.
And so a fourth test of the Jesuits' ability to predict an eclipse was
made. A partial solar eclipse would be visible from China on September 1. Schreck and the Jesuits would produce their prediction.
And the Manchu leaders appointed a former Ming commander, Yang
Guangxian, who was now posing as an astronomer, to produce a
separate prediction based on traditional Chinese methods.

Yang had a checkered past. He had been born in a remote territory in southern China. As a young man he had come to Beijing
where at first he earned a living by blackmailing people. On one
occasion he publicly accused Ming officials of corruption and
incompetence. In defiance of their authority, he had his own coffin
built and dared them to arrest and execute him. Unable to tolerate
his accusations, Ming officials did arrest him. He was convicted
and sentenced to jail and execution. But, soon after his convic-
tion, the Ming Dynasty fell and Yang was freed. How he found
his way to the Bureau of Astronomy and Calendars is not known.
But once there, he became a vocal critic of the Jesuits, calling for
their expulsion from China. It was probably his outspokenness that
brought him to the attention of the Manchu leaders.

Eclipse predictions were made by Schall and by Yang. On the
day of the event, Schall added another element. He produced a
box with a pinhole so that the beginning and the progress of the

partial eclipse could be watched easily.* Now it was a matter of the time of day of the eclipse. Yang had missed it by more than an hour. Schreck, again relying on the *Rudolphine Tables*, came within a few minutes.

After the eclipse, the emperor issued an edict that directed the Bureau to produce a new calendar "in accordance with the Western methods." He also appointed Schall a member of the Bureau and removed Yang and his supporters from the Bureau. Yang was then banished from Beijing.

A year later, on November 19, 1645, Schall presented a new calendar to the emperor. The ceremony was conducted outside the Meridian Gate at the southern entrance to the Forbidden City. All who attended were in formal Chinese attire. All the men, including Schall and the other Jesuits, had the standard Manchu hairstyle—most of the head shaved, leaving a long queue in back. A large table covered with yellow silk was set up in the middle of the Hall of Supreme Harmony in the very center of the Forbidden City. Schall handed the books that contained the new calendar to two high-ranking Manchu officials who carried them to the table and presented the books to the emperor. The two officials then knelt down and kowtowed three times, then backed out. A royal announcer then proclaimed, "In the second year of the Shunzhi emperor's reign, this temporal calendar is bestowed on you." Then everyone present, except the emperor and members of his family, knelt, then pressed their foreheads to the stone pavement nine times, rose, then knelt again, repeating the action thrice. Then the ceremony ended.

---

* The method of watching the progress of a solar eclipse through a pinhole has been known since antiquity. In the fourth century B.C.E. Aristotle called attention to the fact that sunlight passing through a sieve or through leaves of a tree produced crescent-shaped figures. In 990 C.E. the Arab astronomer Ibn al-Haytham mentioned how he made holes in the shutters of his windows to watch a solar eclipse. And in 1485 Leonardo da Vinci was in Milan and used a needle to punch holes in a sheet of paper so that he could watch the progress of a partial solar eclipse.

The details of the calendar, and especially how eclipses were predicted, were still a state secret. Only the emperor, his family, high-ranking Manchu officials, and Schall and a few others at the Bureau would have access to it. In fact, it was forbidden for anyone else to have copies of all or part of the calendar books or to possess any device that might be used to observe the heavens. The minimum punishment was one hundred strokes with a heavy bamboo. In some cases, a person who had unauthorized possession of the calendar or an observing device could be sentenced to death.

For nearly twenty years, Schall ran the Bureau for the Qing Dynasty. He also advised on the casting of cannon and the computation of commercial transactions. In 1661 Emperor Shunzhi died of smallpox. He was twenty-three. His son, Kangxi, then became emperor. He was seven. And so a council of four regents took charge of China. All four were against the Jesuits and ordered most of them out of Beijing. Schall and his closest adviser were put on trial for treason. The trial lasted seven months.

Yang was brought back to Beijing to prosecute the case against Schall. He charged the Jesuit with plotting against the legitimacy of the Qing Dynasty by promoting his Christian religion. As proof, Yang offered thirty churches that had been built in the Qing empire's most strategic locations and contended that these churches harbored foreigners who had secretly entered China to overthrow the dynasty. Then Yang made his most devastating claim. Years earlier, the emperor's concubine, Xian Rui, of noble Manchu birth who had become empress, had given birth to a son who lived only a few months. He was buried on a date selected by Schall as part of his duties at the Bureau. Two years later, Xian Rui died. Her unexpected death, so Yang claimed, was caused by her son's burial on an inauspicious day, a day selected by Schall, which showed that the Jesuits were undermining the Qing Dynasty.

The case was heard at the Ministry of Rites. Its members agreed with Yang, and the case was sent to the Ministry of Punishment for sentencing. The initial sentence was to kill each Jesuit, including Schall, by strangulation. Then the judges at the Ministry of Punishment reconsidered and changed the sentence to death by dismemberment. For two months Schall and other Jesuits were held in jail, tied in such a way they could neither stand nor sit. By now, a new Jesuit, Ferdinand Verbiest, had arrived from The Netherlands. He had expected to assist Schall at the Bureau. Instead, Verbiest spent his time comforting Schall and the other Jesuits who were expecting to die soon.

But fate intervened. On April 15, 1665, the eve of when the death sentence was to be carried out, a strong earthquake struck Beijing. Many buildings collapsed, including the one where Schall and the others were awaiting execution. The emperor saw this as a sign, and he had Schall and the other Jesuits released, though five of their Chinese assistants who had converted to Christianity were still executed. Schall died a year later, weakened by the suffering he had endured during his confinement.

A year after his death, the emperor Kangxi, though only thirteen years old, took power. He had one of the regents who were running the government murdered, and purged the other three. From his interaction with Verbiest, who had succeeded Schall as a leader of the Jesuits in China and who had been teaching the emperor mathematics, Kangxi concluded that the Jesuits posed neither a political nor a military threat to him. He also realized that he needed a Bureau of Astronomy and Calendars that could predict eclipses accurately. But what to do about Yang, who was now in charge of the Bureau? The emperor decided that there would be a contest.

He sent messages to both Verbiest and Yang stating that each of them would be given three challenges to demonstrate their predictive powers. The first was to predict the length of the shadow of a gnomon, a large upright stone that served as the centerpiece of a large sundial in the Forbidden City, at noontime on a date chosen by the emperor. The emperor chose December 27, 1668. Both Verbiest and Yang made their predictions. Kangxi sent eight ministers to observe the test. All eight reported that Verbiest had won the first challenge.

The second test was to predict the angular separation between the Moon and the bright star Arcturus on a night chosen by the emperor. This time he chose February 18, 1669. Again Verbiest won the challenge. But the most important one was next.

The calendar indicated a partial solar eclipse would occur on April 30, 1669. Kangxi asked the two astronomers to predict when the Moon's silhouette would first start to cover the Sun. Each man made his prediction. When the day came, they watched.

Yang had predicted the eclipse would begin before noon, but noon came and no silhouette had yet appeared. Verbiest predicted the eclipse would begin an hour after noon. That came and no silhouette had yet appeared. Finally, a few minutes after one in the afternoon, the dark edge of the Moon could be seen to be crossing in front of the Sun. Verbiest had won the third and final challenge.

In view of Yang's many past transgressions, Kangxi at first sentenced the former head of the Bureau to death, but then he commuted the sentence to banishment. Yang left Beijing and headed for his boyhood home in southern China. He died along the way.

For his successes, Verbiest was appointed the director of the Bureau, a post that would be held by a Jesuit priest until 1773 when the Vatican began to suppress members of the Society of Jesus.

Through the years, the emperor and the Jesuit astronomer became close friends. In 1675 Kangxi visited Verbiest's house and gave him a large sheet of paper with a personal message written on it that read: "I revere the heavens." It was in two large Chinese characters. Verbiest had copies of it affixed to Jesuit churches across China. In 1678 Verbiest presented Kangxi with a 2,000-year calendar with eclipse predictions. Called the Eternal Calendar, it was to symbolize the longevity of the Qing Dynasty. Verbiest lived another ten years. Kangxi died in 1722. He was the longest-serving emperor in Chinese history, his reign lasting fifty-two years.

There is a twist to this already remarkable story about confrontation and survival. The Qing Dynasty, the last dynasty of China, lasted until 1912 when the first Republic of China was established. Throughout most of its nearly three hundred years of existence, the calendars that were produced there were based on Kepler's *Rudolphine Tables*, that were based on a sun-centered model of the solar system. The Jesuits who constructed those calendars never acknowledged the truth of the Copernican system. They were forever obedient to the Vatican and the pope, and taught and wrote about the earth-centered Tychonic system.

And while Verbiest and Yang were confronting each other on the three tests given to them by the emperor, the next great advance in understanding the motion of the Moon and in predicting eclipses was already underway in Europe. No longer would the Moon's motion and eclipse prediction be based on searching for lunar cycles. Instead, the first scientific theory of the Moon's motion was being developed.

Enter Isaac Newton.

CHAPTER SIX

A Request to the Curious

This Moment was determinable with great nicety, the Sun's light being extinguish'd at once.

> —Edmond Halley, comment about a total solar eclipse seen from London in 1715

Anyone who reads one of the many excellent biographies about Isaac Newton soon has a collection of interesting stories to share. This is my favorite one.

Newton was known to be fond of cats. A visitor who came to see Newton in his apartment noticed that there were two saucers sitting on a table. Each one was filled with milk. There was a large saucer and a small one. When the visitor inquired why two, Newton answered that he had *two* cats, an adult and a kitten. The visitor then noticed that two holes had been cut into the base of the entry door, a large one and a small one. When the visitor inquired about these, Newton, looking a bit annoyed, gave the same answer. He said he had two cats, an adult and a kitten.[*]

Newton's annus mirabilis, that is, his "miracle year," was 1666, two years before Verbiest and Yang began their astronomical contest in China. His "miracle year" was actually eighteen months during which he made three revolutionary advancements. He developed the "calculus," a mathematical tool that related position, speed, and acceleration, which would be at the heart of developing a theory of motion in the future. He showed how the spectrum of colors that was produced by sending a beam of white light through a glass prism could be reconstituted back into a beam of white light, proving that objects did not generate color themselves, but that color was a property of white light, a point that had been debated for years—and from which would grow the modern theory of optics. And finally, he proposed a theory of gravity in which a gravitational force was directly proportional to the product of two masses and inversely proportional to the square of the distances between them. Again, this point had been debated for years.

But Newton was a recluse of the highest order, seldom interacting with others and having few close friends. He also seldom sent out notices about what he had discovered. And when he did, he avoided discussing what his discoveries meant. And so the three major advancements he made in 1666 remained unknown for almost twenty years. That is until August 1684 when Edmond Halley made his first visit to Trinity College in Cambridge to meet Isaac Newton.

---

[*] Whether this story is true or false, it is known that people who visited Newton's former apartment long after his death did attest to the fact that there were two plugged holes at the base of the entry door of the proper dimensions to allow for the egress of an adult cat and a kitten.

Halley was then twenty-six years old; Newton was forty-one. Halley was then an assistant at the Royal Observatory at Greenwich. He had just published a catalogue of stars in the southern sky he had observed from the island of St. Helena in the southern Atlantic. On that trip, which had lasted almost two years, he had taken several simple scientific instruments, including a barometer, which he used to show that there was a relationship between atmospheric pressure and height above sea level.

When he returned to London, he established close connections with many of the most learned men of the city. Before seeing Newton, he had had informal discussions with the English architect Christopher Wren and the natural philosopher Robert Hooke about the possibility that some mysterious, unseen force was keeping the planets in orbit around the Sun. The three men had agreed that if such a force existed, the magnitude of the force must decrease as the square of the distance away from the Sun. The three men wondered what Newton thought, knowing he was a strong proponent of the idea that mysterious forces existed in nature. Because neither Wren nor Hooke was on good terms with Newton, Halley volunteered to visit the temperamental professor.

Halley arrived unannounced. Newton, knowing Halley by reputation, let him in. After some time, Halley asked him what curve a planet would follow around the Sun, supposing the force of attraction to be the reciprocal of the square of the distance. Newton answered immediately.

"An ellipse."

"How do you know?" asked Halley.

"Because I have calculated it," Newton replied. He rummaged around trying to find the paper on which he had done the calculation, but was unable to locate it. He promised he would redo the calculation and send it to Halley.

Three months later, Halley received more than he expected. Newton had sent him a small treatise of nine pages entitled *On the Motion of Bodies in an Orbit*. Within it was included a demonstration that an elliptical orbit is produced by an inverse-square force. And there was more.

Newton had also shown how an inverse-square force combined with the conservation of angular motion logically led to Kepler's conclusion about the precise way the speed of a planet varied along its orbit. Newton also showed how an additional conclusion by Kepler that related the size of planetary orbit and its period to the mass of Sun could also be determined.

Halley went back to see Newton and persuaded him to write a single long treatise that detailed all of his work about motion under an inverse-square force, now called gravity. The result was *Principia Mathematica*, published in 1687. It is a monumental work, highly regarded not only by scientists and mathematicians, but also by philosophers and historians for its impact on how people changed their perception of nature, adopting a more mechanistic view of the world and our place in it. It is truly one of the most influential books ever written.

Newton's *Principia* developed the fundamental methods of calculus and showed how Kepler's laws of planetary motion were a logical outcome of an inverse-square force, that is, a theory of gravity. It also demonstrated how those same laws could predict the motions of the moons of Jupiter and Saturn that had recently been discovered by use of early telescopes. It correctly explained the cause of ocean tides. And it predicted the Earth's spin would produce an equatorial bulge on the Earth. But one thing it did not do was give a theory of the Moon's motion.

Newton was well aware that his theory of gravity would have to explain how the Moon moved across the sky. He also knew its motion was quite complicated, more so than for other celestial bodies, because it was near the Earth and because, though the Sun was at a great distance, its great mass meant it also controlled the Moon's motion. He and Halley were also aware that an accurate description of the Moon's motion was not merely an academic exercise. If a precise prescription could be determined, ship captains would be able to use it to determine longitude at sea, making navigation safer and more efficient.

Newton labored for several years. In 1702 he published his result, *A Theory of the Moon's Motion*. He showed how the Sun's gravitational attraction caused the Moon's orbit to deviate from a simple ellipse, causing the lunar nodes to move along the ecliptic and the inclination of the Moon's orbit with respect to the ecliptic to change. He also explained small changes in the motion long known to astronomers as "the variation" and "the evection." Halley then took Newton's work and began to compute new lunar tables. He also did something else: He computed the path the Moon's shadow would soon follow across England during a solar eclipse. It was the first time a theory of gravity, and not merely the repetitive nature of lunar and solar cycles, was used to predict a solar eclipse.

Anticipation of the forthcoming 1715 eclipse soon grew into a frenzy. There were the expected dire predictions. Charles Leadbetter who offered himself as a tutor in surveying, navigation, and "Vulgar and Decimal Arithmetick" provided some of the direst. He saw the coming eclipse as leading to a "Frustration of Hopes" and a "Loss of Friends." There would be a scarcity of "Corn and Fruit." The air would become "unwholesome" and the weather "thick and foggy." There would be the "death of Great Cattle and a Mortality of Old People."

And there were those who saw the eclipse as a way to make money. William Whiston, formerly of Trinity College, offered a device, "The Copernicus," of his own invention. He advertised it as a "universal Astronomical Instrument for the easy Calculation and Exhibition of Eclipses, and of all Celestial Motions." It consisted of ten metallic rings fitted to revolve one within another. Small pins were provided to hold the rings in place. Whiston offered it for sale and gave instructions on how to use it at Will's Coffee House near Covent Garden, then, when that establishment went out of fashion, at Button's just up the street. One of the buyers may have been Alexander Pope, who mentioned in his famous diary "the revolutions of eclipses" taught to him by Whiston.

Halley produced a detailed map that showed the oval shape the Moon's shadow would have at the instant it hung over the center of England, while also showing the path the oval would trace as it moved out from the Atlantic and toward the North Sea. Halley was concerned about the negative impression the public might have about the coming eclipse, and so he had a set of broadsides printed from copper plates that reproduced his map. Beneath the map was an account of what to expect.

> The like Eclipse having not for many Ages been seen in the Southern Parts of Great Britain, I thought it not improper to give the Publick an Account thereof, that the suddain darkness, wherein the Starrs will be visible about the Sun, may give no surprize to the People, who would, if unadvertized, be apt to look upon it as Ominous, and to Interpret it as portending evill to our Sovereign Lord King George and his Government.

In the description he also provided a prediction. In London, which would be near the southern limit of the shadow, the middle of the eclipse should happen at thirteen minutes past nine in the morning. He also sent out "A Request to the Curious to observe what they could about it, but more especially to note the Time of Continuous total Darkness," which, he said, required "no other Instrument than a *Pendulum Clock* with which most Persons are furnished."

As a member of the Royal Society, Halley was asked to organize observations of the eclipse from the rooftop the Society's headquarters at Crane Court. Both members and invited guests, including "several foreign Gentleman" who had traveled from the mainland of Europe, were present.

For his observations, Halley procured a quadrant of nearly thirty inches in radius and "exceedingly fixt with Telescope Sights and moved with Screws so as to follow the Sun with great Nicety." He also set up a pendulum clock "well adjusted to the mean Time."

Early on the morning of the eclipse, he pointed the quadrant at the Sun and kept his eye to the telescopic eyepiece. At six minutes after eight o'clock, he "began to perceive a small Depression made in the Sun's Western Limb." This was the first sighting of the Moon's silhouette. It immediately became more conspicuous. At nine o'clock, the darkening sky had turned from an azure blue to a dusky purple. Nine minutes later, the Sun was completely hidden.

"During the whole time of the Total Eclipse," Halley would write,

> I kept my Telescope constantly fixt on the Moon, in order to observe what might occur in this uncommon Appearance: and I found that there were perpetual Flashes or Coruscations of Light, which seemed for a Moment to dart out from Behind the Moon, now here, now there, on all Sides.

When the Sun reemerged, it "came out in an Instant with so much Lustre that it surprised the Beholders, and in a Moment restored the Day."

By keeping one eye on the carefully set pendulum clock and the other to the telescope eyepiece, Halley determined that totality began at nine minutes and three seconds after the hour and ended at twelve minutes and twenty-six seconds. That meant the all-important middle of the eclipse had occurred at ten minutes forty-four seconds, just two minutes earlier than Halley's predicted time.

He gathered the records of other observers scattered across England, determining who had seen totality and for how long. From that, he constructed a second map, one that showed the actual shadow of the Moon as it passed over England. He compared the second map to his predicted map. The predicted map showed a shadow width of 160 miles. The actual width was 183 miles. The southern edge of the actual shadow was twenty miles south of the predicted one; the north edge was only three miles north of the predicted one.

By any measure, it was a notable achievement. For the first time, a position and a width of the Moon's shadow had been predicted—and confirmed. And it had been done with a new theory that postulated the existence of a mysterious force—gravity—that permeated all of space and that influenced every object. Added to this, a new and not-yet-widely-accepted type of mathematics—calculus—had been the basis for the prediction. Even Kepler, just a generation or so earlier, at the end of the preface to the *Rudolphine Tables*, had suggested that there were some motions that the Moon made that seemed to be due to chance. And, as others thought, it was these subtle motions of chance that was evidence that God was still involved in modifying the universe. But a theory of gravity and the logical foundation of *Principa* would eventually dispel this worldview.

Halley continued to apply mathematics to natural phenomenon. His most famous application, which was done years before the 1715 eclipse, led him to realize that the comets sighted in the years 1456, 1531, 1607, and 1682 had similar elliptical orbits. From that, he suggested that they were probably the same comet. And he made a prediction: The comet should reappear in 1758, a year, unfortunately, that was so far in the future that it would be outside his lifetime. But he was confident it would reappear.

Others were not so confident. They thought, as Kepler had suggested, that some phenomena occurred by chance, thus limiting how precisely an event, such as an eclipse or the appearance of a comet, could be predicted. One of these advocates was the Irish satirist Jonathan Swift.

In Part III of *Gulliver's Travels*, published in the years immediately after Halley made his prediction about a returning comet, Gulliver arrives at Laputa, a kingdom in the sky populated by mathematicians. According to their theories, they should be able to extract sunshine from cucumbers and build houses from the roof down. One of the characters always has his eyes on the sky and through mathematics has predicted the return of a comet that will destroy Laputa and the rest of the Earth during its next passage.

In *Gulliver's Travels*, Swift mocks those who rely on calculations, equating their work with the craft of a foolish tailor who measures customers with astronomical instruments. The tailor uses a quadrant to determine a customer's height and a compass to outline his body. The clothes he sews are "very ill made, and quite out of shape, by happening to mistake a figure in the calculation." Finding such a world intolerable, Gulliver leaves.

There was no good way for Swift and Halley to resolve their different worldviews. In fact, the first major test of Newton's theory of gravity would not come until more than a decade after both men had died and it would be based on detailed calculations that predicted exactly when Halley's Comet would return and make its closest approach to the Sun.

Most of the actual calculations were done by Nicole-Reine Lepaute who worked in her apartment in Paris, guided by French mathematician and astronomer Alexis Claude Clairaut. She began the work in June 1757. On November 14, 1757, Clairaut presented the results to the *Académie des sciences* in Paris. The comet, he predicted, would make its closest approach to the Sun between March 15 and May 15 of the next year. The comet was first sighted on its return barely a month later on Christmas Day 1757. It passed closest to the Sun on March 13, just two days earlier than predicted.

———•———

Notwithstanding the successful application of Newton's theory of gravity to predict the return of a comet, during the decades leading up to Clairaut's announcement there was still a concern among some of Newton's contemporaries that invoking a mysterious force that influenced everything was a giant step backward. God was being swapped for gravity.

Halley of course never waivered in his support. The French philosopher Voltaire was immediately enthusiastic and promoted the story that Newton had been inspired after seeing an apple fall from a tree. At the same time, the Dutch mathematician Christiaan Huygens was astonished that a theory had been proposed that relied on a force that acted on everything in the universe equally. Along a similar line, Gottfried Leibniz, who contended with Newton over which of them had invented calculus, accused Newton of abandoning causation and replacing it with an endless series of miracles. In short, Huygens and Leibniz considered Newton's theory of gravity to be a return to believing in the occult.

And there was merit to what they said. In addition to experiments in optics and to the development of the calculus and a theory of gravity, Newton also did experiments in alchemy and delved into theology. He claimed there were hidden messages in the book of Revelation and in the measurements of Solomon's temple. In 1713, in a brief addendum to his work, he suggested that a subtle spirit that pervaded all bodies was the cause of gravitation.

And, yet the theory worked. When questioned why, Newton gave a simple though evasive answer: "It is enough to know that gravity exists."

For his part, Halley realized there was a practical aspect to Newton's theory, one that would solve a problem of great importance.

———•———

In the fall of 1707, five years after Newton proposed his lunar theory, a fleet of twenty-one British ships under the command of Sir Cloudesley Shovell was returning to England after engaging a French fleet in the Mediterranean. The passage back was marked by bad weather and constant gales. On the night of November 2, Shovell's sailing masters informed him they were in safe waters west of Ushant, the western most point of France, far south of the British Isles, and that only the open Atlantic Ocean lay to the west of them. Or so they thought. But they were wrong. As the fleet tacked west, it was closing in on the Scilly Islands.

Several ships struck rocks, and four were lost. The number of dead is unknown, but was probably in the thousands. The wrecking of part of the fleet and the loss of so many lives so near to home raised a concern. Though the cause was a mistake in latitude, not longitude, members of the British Admiralty called on the government to support an improvement in navigation. In 1714 the British Parliament passed the Longitude Act.

The act established the Board of Longitude that offered monetary awards to those who could improve the determination of longitude at sea. Those who sought the awards generally followed one of two basic strategies. In one strategy they either built or improved mechanical clocks so that they would endure the rough passage of a sea voyage and still provide an accurate time. Here the main character was John Harrison who designed and built four clocks, or chronometers, for the Longitude Board. The most important one, known as H4, which he called "the Sea Watch," was about five inches in diameter and was powered by the winding of a coiled spring. It was the forerunner of the later pocket watch and, much later, the first popular wristwatches. The other strategy that was followed relied on a misleadingly named technique known as "the method of lunar distance."

The method of lunar distance relied on measuring the angular separation between the Moon and another celestial object, usually a bright star. Because the Moon moves with respect to the stars, if one also had a table of accurate positions of the Moon, then one would merely compare the angular measurement with the table to determine local time and, from that, the longitude.

In principle, the method of lunar distance was easy. In practice, it was difficult because it required an accurate theory of the Moon's motion. According to the Longitude Act, to qualify for an award, one had to determine longitude within $1°$ on the Earth's surface, which, along the equator, equaled about seventy miles and, at the latitude of London, about forty-five miles. For the method of lunar distance to work, the Moon's position had to be predicted to within about $\frac{1}{30}$ of that amount, or $0.03°$ on the sky. As Halley knew, the lunar theory that Newton proposed in 1702 could locate the Moon to only 0.15°. And so another major improvement was needed.*

An improvement did not come until a decade after Halley's death in 1742. That is not to say that the intervening years were devoid of progress. The greatest mathematicians of the age worked on improving Newton's lunar theory. There was Leonard Euler who, in 1735, had lost one eye, a disaster that caused him to remark the he would now be less distracted from his mathematics.** To lunar theory, he added a method that used what are today familiar trigonometric functions, such as sine and cosine. In France, Clairaut, who would supervise the calculations that predicted when Halley's Comet would return, and Jean le Rond d'Alembert were rivals. At first each questioned Newton's work, then, satisfied that an inverse-square law for gravity was correct—and ignoring the criticism about an all-permeating mysterious force—competed to account for slight deviations in the Moon's motion from an elliptical orbit. But the next major advance did not come from a mathematician or from an astronomer. It came from a German mapmaker.

Tobias Mayer was born near Stuttgart, Germany, in 1723. When Tobias was nine years old, his father died. His mother, unable to support him and his many brothers and sisters, sent the youngest ones (including Tobias) away to live in what were essentially foster homes. It was in one such home that Mayer became acquainted with a shoemaker who had interests that were much broader than his occupation might suggest. In particular, the shoemaker was passionate about mathematics. He had money to purchase books on

---

* For comparison, the Moon's diameter is 0.5°, about half the width of your thumb when held at arm's length. Also, the reason the Moon's position had to be known to 0.03°, which is 30 times smaller in angular measure than the 1° of longitude, is because it takes 30 times longer (about 30 days) for the Moon to orbit the Earth than for the Earth to turn once on its axis, which is one day. That is, when the Earth rotates 1° in longitude, the Moon has moved only 1/30th that amount in angular measurement in the sky, 0.03°. Think about it.

** Seventeen years later, Euler lost sight in his other eye. Though now totally blind, he continued to improve the mathematical theory of lunar motion.

the subject, but never had the time to read them. So young Mayer took on the task of reading the books, then discussing the contents with the shoemaker.

As a young man, he found work at a cartographic shop copying maps. He became interested in how geographical information was compiled to make the maps. This is where his self-education in mathematics paid off. He was surprised to learn that in the hundred or so years since James and Gellibrand had measured the distance across the Atlantic from England to Hudson Bay, only about an additional hundred good measurements had been made. And that was because of an inaccurate lunar theory.

He studied the theories of Newton and others, especially, Euler. He built his own telescope so that he could record the precise moment when the Moon passed in front of a star, thereby making the most exact measurement of the Moon's position in the sky. He combined those measurements with the current theory of lunar motion, and in 1753 produced his own lunar tables. He submitted them to the Longitude Board. Members of the Board were amazed: Mayer's tables were superior to any then available. They also passed the necessary test of predicting the Moon's position to better than 0.03°. He was given one of the Board's monetary awards.

Mayer's work was the basis for the lunar tables in the *Nautical Almanac*, which began to be published annually in 1767. A copy could be found on every ship of the British fleet. Captain James Cook carried one of the first copies when he sailed from England in 1768 on his first voyage to the Pacific Ocean. Mayer's tables proved valuable to Cook and to other voyagers, allowing them to determine longitudes in remote parts of the world.

—·—

In the span of a little more than fifty years since the 1715 eclipse and the fervor Halley's mathematical prediction and map created, a shift in how eclipses were regarded began to take place. Whiston, who lectured about the 1715 eclipse alongside Halley and sold a device, which he called "The Copernicus," to aid people in following it, said later that "the sale of my schemes before and after" earned enough money to support his family for a year. Another total solar eclipse passed over England in 1724, ending at sunset as seen from Europe. Both Halley and Whiston published maps for that event, further promoting an interest in solar eclipses whereas before there had been mostly mystery, fear, and dread.

People now looked forward to eclipses. An annular solar eclipse passed over Scotland and the southern parts of Norway and Sweden in 1737, and people in the towns and cities of those countries made a concerted effort to view the eclipse and record what they saw. Eleven years later another annular eclipse passed over Scotland, though this time the path headed south crossing over Denmark, northern Germany, and Poland. In 1753 the very narrow path of a total solar eclipse crossed through northern Portugal and central Spain. In 1764 people in Portugal, Spain, France, Belgium, the Netherlands, Germany, Denmark, Norway, Sweden, and Finland all saw the narrow ring of sunlight that remained in the sky during an annular eclipse. In England, the Moon was seen to pass directly in front of the Sun from Ipswich and Canterbury, but not farther to the west in London or at the breeding stables of thoroughbred racehorses at Windsor Great Park where a foal was born on the day of the eclipse, April 1, and was appropriately named Eclipse. Eclipse would compete in eighteen races, and he won them all. He was retired to stud and is in the pedigree of more than 90 percent of the thoroughbred horses racing today.*

Eclipse watching was also taken up across the Atlantic in the American colonies. In 1684 Harvard President Increase Mather delayed commencement ten days so that faculty, students, and other interested persons could travel to Martha's Vineyard to see a total solar eclipse.

---

* Contrary to popular belief, the Eclipse automobile, first manufactured by Mitsubishi in 1989, was named for the racehorse and not for the natural phenomenon.

Decades earlier, in 1643, Benjamin Franklin used timings of a lunar eclipse to come to a remarkable conclusion about hurricanes. It was the evening of October 21, and he was preparing to watch a lunar eclipse that, by checking his own almanac, he knew would start at about 8:30. Unfortunately, a violent storm hit Philadelphia that night, blanketing the sky with clouds for days. "But what surprised me," he later wrote to a friend, "was to find in the Boston newspapers an account of the observation of that eclipse." But how was that possible? In Philadelphia, the wind had come from the northeast. Wasn't that the direction that the storm had come? Franklin sent inquires up and down the coast, discovering, from the timings of the eclipse, that the storm had moved to the northeast and that the wind associated with it blew as a counterclockwise rotating vortex. It was the first suggestion of the true nature of hurricanes.*

After 1684, the next solar eclipse to pass over the northern half of the colonies was in 1778, during the Revolutionary War. David Rittenhouse, who was then treasurer of the state of Pennsylvania and a noted astronomer, attempted to view the eclipse from Philadelphia just eight days after the British army had evacuated the city. It was clouds, and not the British, that prevented him from seeing the eclipse. Thomas Jefferson hoped to use the eclipse to determine the longitude of Monticello, but clouds prevented him from making any measurements. Among the few people who reported seeing the 1778 total solar eclipse was Antonio de Ulloa, admiral of a fleet of ships sailing between the Azores and Portugal. He had an unobstructed view of the skies and saw totality for four minutes, though the motion of the ship delayed setting up telescopes and seeing the entire eclipse.

---

* Another unusual use of a lunar eclipse was made June 27, 1927, when Edison Pettit and Seth Nicholson at Mount Wilson Observatory in California measured the rate the lunar surface cooled when the Moon entered the Earth's shadow. From that, they concluded that a layer of dust must cover most of the Moon. Decades later, the first spacecraft landed on the Moon and confirmed their conclusion.

The next opportunity came on October 27, 1780, also during the Revolutionary War. Samuel Williams, who had recently been appointed a professor of mathematics at Harvard, used Mayer's lunar tables to determine that the path of the total solar eclipse would run across Canada and exit North America at Penobscot Bay, which was then under the control of British forces.

Williams wrote to John Hancock, Speaker of the Massachusetts House of Representatives, asking if he would intercede on Williams's behalf and request safe passage for him to view the eclipse. Hancock did so, sending a personal message to the British commander at Penobscot Bay. "Sir. It is expected there will be a very remarkable Eclipse of the Sun . . . that will be central & total at or near the British Post at Penobscot where you command." Hancock asked the commander to prove himself a "Friend of Science" and grant safe passage for Williams and a few others in a single ship to set up an observation site. "Though we are politically enemies," Hancock wrote, "yet with regard to Science it is presumable we shall not dissent from the practice of all civilized people in promoting it wither in conjunction or separately."

The commander granted the request, provided that Williams and whoever traveled with him limit their stay to five days.

Williams and seven others set off in the sloop *Lincoln*. They arrived on the island of Islesboro about four miles off the west shore of Penobscot Bay. Islesboro is about the size of Manhattan, though narrower. Williams and the others set up camp near the middle of the island. On eclipse day, they watched with what must have been great anticipation as the Moon slowly passed in front of the Sun. Eventually, there was a crescent of brilliant sunlight left, then just a sliver. And the sliver continued to narrow. Williams must have felt that he was just seconds away from seeing a total eclipse when the unthinkable happened. The sliver started to widen and the crescent reappeared. Then the Moon moved entirely off the Sun. What had gone wrong?

Williams would maintain the map he used showed the wrong latitude. But modern scholars have checked the map he used and find no problems. Perhaps Mayer's tables were in error. Again details of the 1778 solar eclipse have been recalculated using Mayer's tables, and no errors in the tables have been found. Which leads to the inevitable conclusion: Williams must have made a mistake in his eclipse calculation.

After months of planning and preparation and after being fortunate enough to have clear weather, the last thing any eclipse watcher wants to admit is that the prediction was wrong. But that seems to be the case for Williams in 1778.

In his defense, eclipse calculations, even with Newton and Halley's new methods, were still tedious and laborious, requiring hours of work. An easier method with fewer calculations, and so, less potential for an error, needed to be found.

———•———

You have two pans that you use to bake cornbread. One is square and the other is round. You have sixteen people who are coming to visit and each one expects an equal share of the cornbread. How do you cut the cornbread to make sure everyone gets an equal share?

For the square pan, it is easy. You make three cuts at equal intervals that are parallel to the edge of the square pan. Then you rotate the pan a quarter turn—that is, by 90°—then make three more parallel cuts at equal intervals. The result is sixteen equal pieces of cornbread from the square pan. But how do you cut pieces from the round pan?

You start by making a single cut along any diameter followed by turning the pan 90° and making a second cut along a diameter. What do you do now if you follow the same rules used in cutting the square pan? If you follow the same rules used in cutting the square pan—that is, all cuts must be either parallel or perpendicular to the first cut—then the problem is very difficult, but it can be done, and the pieces will all have the same amount of cake but they will look strange because of the curved edge of the round pan. An easier way is to make two more cuts along diagonals to produce sixteen pieces of cornbread, each one pie-shaped and equal in size. This is to illustrate that if you take advantage of an unusual geometry, in this case, a curved edge, your life can be much simpler.

In essence, Friedrich Wilhelm Bessel did the same thing for the calculation of eclipses.

Since the days of Mayer, Halley, Kepler, and earlier mathematicians, calculations for a solar eclipse were based on chasing the Sun and the Moon across the sky to see when and where they met. Bessel changed the perspective. Instead of following the Sun and the Moon across the sky, he asked: When does the Moon's shadow touch the Earth and, when it does, what shape does the shadow have on the Earth's curved surface?

Admittedly, Bessel's approach does use spherical geometry, which is complex but—believe me—it is much easier to calculate solar eclipses this way versus chasing eclipses individually and making measurements by hand every time. There are fewer potential errors to make using Bessel's calculation, unlike the one or more that Samuel Williams apparently made using Halley's procedure.

Bessel's method is used still today. For a given eclipse, eight parameters, appropriately called Besselian elements, are computed. Today they are usually tabulated at hourly intervals. It is from these sets of Besselian elements that maps are made that show the path of totality, where on the Earth's surface a partial eclipse will be visible, where the eclipse will be seen at sunrise or sunset and so forth.

Bessel introduced his new method in 1824. Thus the stage was set. Mayer's tables and those that evolved from it yielded eclipse predictions to the nearest minute. Where one needed to stand to see totality could be determined to within a few miles. People could now plan years in advance where they would have to be to stand within the Moon's shadow.

And as they made their journeys, sometimes to the far reaches of the Earth, the same thought ran through each person's mind: I wonder what I will see.

CHAPTER SEVEN

The Annulus at Inch Bonney

How can I express the darkness? It was a sudden plunge, when one did not expect it.

— Virginia Woolf, 1927 solar eclipse

D uring the first decades of the nineteenth century, one of the most highly regarded and highly successful members of the London Stock Exchange was a seller of annuities named Francis Baily. He wrote numerous pamphlets on the subject, explaining to prospective clients how annuities worked and how to invest in them. Both he and his clients became wealthy. And he would probably have continued in that business if his faith in the essential honesty of others had not been shaken by a giant fraud.

It was 1814 and a British staff officer had brought news from the Continent that Napoleon had been defeated and killed by members of his own army. The stock market in London soared. Then those who had arranged for the false news sold their funds before it was discovered that Napoleon was still alive and still engaged against the British Army. Baily conducted his own investigation in secret and uncovered those involved in the fraudulent dealings. In fact, he amassed so much evidence against the perpetrators that, as the prosecuting attorneys openly acknowledged, no more complete and conclusive change of evidence had ever before been produced in court. Convictions came quickly. But Baily was now disillusioned. And so, with his considerable wealth, which he had acquired legally, he left London and settled in the nearby countryside to pursue an area of study that had long fascinated him—astronomy. And within the field of astronomy his interest soon centered on eclipses.

He worked, as others had before him, on determining chronologies based on calculating the day and the year of past eclipses. He also did calculations that showed when, in the future, the Moon would pass in front of particular bright stars, and from those calculations produced a long series of detailed charts that he sent to the Foreign Office of the British government, which sent them to overseas stations where they were used to make better determinations of longitude. He also computed the tracks of solar eclipses. And one expects he must have been overwhelmed with excitement when he discovered that the track of the solar eclipse of 1836 would pass close to him. He would be able to see the eclipsed Sun from Scotland.

Baily laid out a map. Using the Besselian elements that were now being published in the *Nautical Almanac*, his calculations showed that the central path of the 1836 solar eclipse would pass from Ayr on the western coast of Scotland to Alnwick on the eastern coast. Baily drew a line connecting the two towns. He studied the line and saw that it passed close to Makerston, Scotland, where Baily knew there was a well-furnished private observatory where he would be able to set the four pocket chronometers that he would be carrying. And south of Makerston, about eight miles, was the home of Scotland's best known designer of scientific instruments, James Veitch, who lived in the quaintly named village of Inch Bonney.

The name "Inch Bonney" is a hybrid of Lowland Scots, *bonnie* meaning "pretty," and of Scottish Gaelic, *inns* meaning island, often anglicized as "inch." Veitch, for all his later notoriety among scientists, was first famous for designing farm ploughs. His were lighter in weight, less expensive, and more efficient in draught. But his true passion was in designing and building delicate scientific instruments, such as barometers, clocks, microscopes, and telescopes. He also fitted his neighbors with customized glass spectacles and toyed with early electrical machines. He and his wife were often found staring at the sky, and when the occasion rose, stopping strangers to point out the planet Venus visible in daylight.

As soon as Baily arrived at Inch Bonney, he made the acquaintance of Veitch. Outside Veitch's house, the men set up the two telescopes Baily had brought to observe the eclipse. Baily had also prepared small glass panes that he had smoked heavily with soot, which he and Vetch would use to look directly at the sun.

On the morning of the 1836 solar eclipse, which would be an annular one, the weather was remarkably clear. Not a cloud could be seen in any part of the sky that morning or later during the eclipse.

The first indication of the Moon covering the Sun was seen at 1:36 that afternoon. It would take one hour and twenty-five minutes for the Moon to completely cover the Sun. With just one minute to go, the still uncovered part of the Sun now a very thin slit, Baily removed the panes of smoked glass that covered the large lenses of both telescopes. He looked through the eyepiece of one. The scene, he would write, so riveted his attention that he could not take his eye away from the telescope to note down anything during the progress of what was a highly unusual and totally unexpected phenomenon.

A row of lucid points, like a string of bright beads, unevenly spaced, suddenly formed along that part of the circumference of the Moon that was about to enter on the Sun's disk and close the slit. The formation of these beads was so rapid that, to Baily, "it presented the appearance of having been caused by the ignition of a fine train of gunpowder." As the Moon pursued its course, the dark intervening spaces stretched out into long black lines, eventually joining the limbs of the Sun and the Moon. At that moment, a complete ring, or annulus, formed around the silhouetted Moon, its edge completely smooth and circular.

He described to Veitch, who was looking through the other telescope, what he had just seen and asked if Veitch had also seen it. He said he had.

The Moon continued to move. Soon the same phenomenon occurred in reverse order on the opposite side of the Moon: Several bright spots formed that merged into a few long streaks, then a dazzling bright sliver as the Sun began to reappear.

Baily later questioned others who had observed the eclipse from other locations. Many of them said they had seen the same thing through their telescopes. The obvious explanation was that the luminous spots were rays of sunlight shooting between mountains and through valleys on the Moon. But Baily realized that could not be the full explanation.

He adduced correctly that what he had seen was similar to the infamous black-drop effect seen at the beginning or the ending of a transit of Venus across the Sun. When Venus first appears on the Sun's limb, the planet's round disk is distorted so that the side that is next to the Sun's limb appears to adhere and a dark protuberance forms. As the planet advances, the protuberance increases in length and narrows in width until the round disk does appear.

It was not until the 1990s that it was shown that this is an optical effect that occurs because the Sun's disk is not uniformly bright. Instead, the disk is noticeably darker close to the limb. It is the reason that the luminous spots seen by Baily and others seemed to dart back and forth then suddenly disappear.

Baily read through accounts of earlier eclipses, deciding that others had already reported the phenomenon. During the 1820 annular eclipse, Jean Henri van Swinden, a mathematician in Amsterdam, had seen "several dark threads, lines, or belts" just before the annulus formed. There were similar reports for eclipses in 1806, 1791, and 1724. The earliest account was the one that Halley made in 1715 when he saw "Flashes or Coruscations of Light, which seemed for a Moment to dart out from behind the Moon, now here, now there, on all Sides."

Today we associate the phenomenon with Baily—the luminous spots are now known as *Baily beads*—because he gave the first vivid description of them. And, as it is remembered, he viewed them through a telescope when the Sun was almost totally obscured by the Moon.

———•———

But it was not *completely* obscured. For that reason, it is important to caution anyone who might try to repeat Baily's effort. His telescopes were inferior in optical quality to telescopes or binoculars that can be purchased today, even inexpensive ones. Though Baily was able to view the luminous spots magnified through his telescope, to do so today would almost certainly harm a person's eyesight.

The ancient Greeks knew of the potential harm of viewing a solar eclipse directly. According to Plato, Socrates warned: "People may injure their eyes by observing or gazing on the sun during an eclipse." More than a thousand years later, the Arab scholar Ahmad al-Biruni repeated the warning: "If a person looks directly at the sun, his sight becomes dazzled and confused." Al-Biruni even admitted that his eyesight was weakened during his youth by staring at solar eclipses. Both Socrates and al-Biruni also knew of a safe way to view the progress of a solar eclipse—by looking at the Sun's diminished image reflected on water. Baily saw the technique used during the 1820 partial solar eclipse that he watched from London. And people continue to use it today.

Newton suffered a temporary loss of vision when he was conducting an experiment on himself by looking at the reflection of the Sun in a mirror while standing in a darkened room. He began by looking at the mirror with one eye, then turned toward a dark corner of the room and blinked to observe how long it took the residual image of the Sun in his eye to disappear. He repeated this, staring at the mirror for longer and longer intervals. Eventually, after several trials, he noticed that the Sun's image never diminished. He spent the next three days in a room too dark to read or write, and yet the Sun's image persisted. Finally, on the fourth day, his eyes began to recover, though he continued to lie in bed at night with curtains drawn.

One who was temporarily blinded by watching the partial phases of a total solar eclipse was Taddeo Gaddi, a student of the famous Italian painter Giotto. Gaddi later wrote to a friend:

> From days not long past I have suffered, and still suffer, from an unendurable infirmity of the eyes, which has been occasioned by my own folly. For while, this year, the Sun was in eclipse I looked at the Sun itself for a long period of time, and hence the infirmity . . . For I constantly have clouds before my eyes which impeded the vigor of my sight.

The eclipse occurred on July 16, 1330. At the time he was completing the fresco *Annunciation to the Shepherds* for the Baroncelli Chapel at the Basilica of Santa Croce, Florence. It is a night scene that shows a shepherd on a hillside. Some art historians have suggested that this muted image illuminated by light coming from the heavens may stem from his partial blindness.

More recently, in 1962, after a partial solar eclipse that was visible from the Hawaiian Islands, there were fifty-two cases of partial blindness, an affliction now known as solar retinopathy. In 1996 in India, twenty-one people who were using so-called protective devices, such as sunglasses with ultraviolet, or UV, protection or exposed X-ray film, had eye damage. In Europe in 1999, seventy people were diagnosed with a loss of vision during a solar eclipse.

Though none of these people went totally blind, some did permanently lose part of their eyesight.

What happens to the eyes when they are exposed to intense sunlight? First, no pain is felt. Second, symptoms may include a blurring or distortion of images or an impairment of color vision. And symptoms may not be apparent for hours or days after exposure. Physically, there is a rise in the temperature of the fluid contained within the eyeball that triggers a change in the photochemistry of receptor cells on the retina. Cells of the macula, the small spot on the retina where vision is keenest, may be damaged permanently by intense light, the exposure leaving lesions. Damage seems to be cumulative. And no treatment has been shown to be effective.

So how does one safely view a solar eclipse? Indirect means are the safest. The intensity of sunlight is greatly diminished after it is reflected off water. The Sun's image can also be projected through a pinhole onto a blank surface. A telescope or binoculars can be used to project an image; again, the image is projected onto a blank surface. The Sun can be viewed directly only when a filter specifically designed for intense light is used. Such filters have a metallic coat that attenuates not only visible light but also harmful ultraviolet and infrared light. Aluminized Mylar sheets have been popular, but, because of their fragility, small cracks or pinholes may develop. The best way, if one must look at the Sun directly, is to glance occasionally through No. 14 welder's glass, available though welding supply stores. Unsafe filters include exposed film, including X-ray film, smoked glass, polarizing filters, or photographic neutral density filters, multiple sunglasses, Mylar blankets designed for warmth on cold nights, and CDs or DVDs. The fact that the Sun appears dark in a filter does not guarantee that it is safe.

During the few minutes, or seconds, when the Moon completely covers the Sun during a total solar eclipse, it is safe to use any means to observe the eclipse, though one must exercise extreme caution to put protective eye covering back in place before the Moon's shadow passes and the blinding Sun suddenly reappears. Nevertheless, as Baily discovered, it was during a total solar eclipse that he was "electrified at the sight of one of the most brilliant and splendid phenomena that can well be imagined."

—·—

The third solar eclipse in Baily's life occurred in 1842 when the Moon's shadow would sweep across Spain, southern France, northern Italy, Austria, and the eastern part of Europe. His original intention was to take up a station in the south of France, but when he arrived with a few days to spare before the eclipse and because he intended to visit Venice on his return, he changed his mind and continued eastward along the route the Moon's shadow would take. He eventually stopped at Pavia in Italy about noon on July 7, the day before the eclipse. Here a professor at the university, having heard of Baily's arrival, offered him the use of any of the apartments at the university. Baily selected one of the upper rooms with a window facing east, through which he would watch the eclipse. The professor asked if Baily needed of any other assistance. He responded that all he needed was the key to the room, then, when the professor left, Baily locked the door to ensure he would not be disturbed during the eclipse.

At sunrise a thin layer of clouds was seen in the east near the horizon, but the Sun soon rose above them and the rest of the sky was clear the remainder of the day. Just before 6:00 A.M., Baily saw the commencement of the eclipse as the Moon began to cover the Sun. Ninety minutes later, he studied the thin crescent of sunlight that remained, anticipating the black lines that would precede a string of luminous beads. The lines never appeared, but the beads were quite distinct. As he counted the number of seconds that the beads were visible, he heard a burst of applause from the streets below. He looked down at the street, and then returned to the sky. At that moment, the dark body of the Moon was now surrounded by an incandescent white glory, similar, Baily would write, "in shape and relative magnitude to that which painters draw round the heads of saints."

I was somewhat surprised and astonished at the splendid scene which now so suddenly burst upon my view. It rivetted my attention so effectually that I quite lost sight of the string of *beads*, which however were not completely closed when this phenomenon first appeared.

This was the *corona*. In breadth, it extended nearly the equal of the Moon's diameter away from the outline of the Moon. Its light was brightest next to the Moon's border and diminished gradually with distance. In color, Baily saw it as "quite white, not pearl color, nor yellow." In anticipation of the darkness that would come with the total solar eclipse, Baily had lit a candle, but discovered it was not needed, the corona producing enough light, by Baily's estimate, to read small print. But, even with the grandeur of the corona in view, Baily, as he continued to look through his telescope, saw something equally marvelous and totally unexpected.

At three places around the border of the Moon were small protuberances. (Today these are known as "prominences.") Each one had a different irregular shape that Baily likened to tall mountains. All were of the same vivid shade of red. Their light was steady and had none of the flickering or sparkling motion visible in parts of the corona. Baily watched them until a new set of luminous beads formed on the side of the Moon, opposite to where the first set had been, signaling the reappearance of the Sun's bright disk and the end of the total eclipse.

A second British astronomer had also made a trip to Italy to see the eclipse. He was George Airy, Astronomer Royal, a title that dated back to 1675 and that Halley once held. For his viewing station, Airy had chosen the Basilica of Superga located on the highest point of a cluster of hills that overlooked the Po Valley and the city of Turin. He had arrived the day before the eclipse and made the climb to the Basilica during the night, waking the fathers who offered him any part of the church to make his observations. High on the church dome seemed to be the natural place, but after a quick survey Airy chose a platform in front of the portico. He set up his telescope in the dark of morning twilight. Then he and the fathers waited for the Sun to rise.

Airy was also favored with clear skies. Two minutes before totality, the sky seemed to darken appreciably. A sliver of sunlight was still visible, and, to see it, Airy squinted his eyes. He was just ready to put his eye to the telescope when a set of Baily beads came into view. He tried to time them with the beating of his heart. After just two beats, the beads disappeared and the corona was revealed. He, too, with his unaided eye, saw three small red protuberances extending out from the Sun. As did the fathers gathered around him, one of them taking the time to study the landscape around them. "The general appearance was very frightful," he would write. "It was like looking at objects through a very dark greenish glass." Airy also noted, as Baily had, that at the moment when the Moon completely covered the Sun, the people standing around him burst into spontaneous applause.

————•————

The accounts of the 1842 solar eclipse by Baily and Airy, while scientifically accurate, do not capture the visceral feelings one has while viewing a total solar eclipse. Fortunately, in 1842, Austrian writer Adalbert Stifter was in Vienna and he witnessed the event.

"Never, ever in my entire life," he wrote, "was I so shaken, from terror and sublimity so shaken, as in these two minutes." It was such a simple thing: the Sun illuminates the Moon, which, in turn, throws its shadow upon the Earth. And, yet, because the bodies are so large and so distant, it felt, to witness one, as if the laws of nature had been disturbed.

On the morning of the eclipse, Stifter awoke early and climbed to the top of a house where he had a view of the entire city of Vienna and the countryside all around. The Sun had just risen. He already had a strange feeling, anticipating what would occur over the next two hours. But the experience was altogether different from what he expected.

He could see that people had gathered atop other buildings. Heads appeared from attic windows. Some joined him at the top of his house. All were turned toward the Sun, waiting for it to go out.

"It was strange," Stifter wrote, "that this uncanny, clump-like, deep black thing approaching us, which slowly ate away the sun, should be our moon."

Then came the first feeling of discomfort as the sky and the landscape grew darker. The change in light was not like that of an evening twilight with reds and greens, but a somber change, colors fading away to shades of gray.

"It's coming" was the shout. The tension now reached a peak. The Sun faded away, like the last spark of a fading ember. Stifter thought it was an extremely sad moment. Then, in spite of the calculations and knowing exactly when the Sun would disappear completely, there was still an irrational feeling of dread that was followed by one of wonder. Stifter watched as a woman near him started a heartfelt cry. Another in a house next to him fainted. And a man later confessed to Stifter that tears had poured from his eyes.

"I have always taken the old descriptions of eclipses to be exaggerated," Stifter later judged, "just as perhaps in a later time this one will be as well; but all of them, like this one, are far from the truth."

Finally, at the moment of totality, a cheer rose from the crowd. Everyone applauded.

The absence of the Sun lasted a short time, less than two minutes. Then a crescent of brilliant sunlight appeared. The crescent grew larger. Normal colors and shadows returned. People returned to their work, though, as Stifter recorded it, for days they would tell each other how they could still feel the vision of the eclipse deep within their hearts.

I had seen two total solar eclipses when I first read Stifter's account, and I was surprised at how accurately he was able to capture, in a way no one had before, the emotions one feels when viewing an eclipse and during the aftermath. To his account I would add just one thing: For me, the sight of a total eclipse—the sudden darkening of the sky, the radiant corona, the blood-red prominences circled around the edge of what had just been a brilliant Sun—was the most primal experience I have ever had. It was as if the most primitive part of the brain—the part inherited from reptiles—kicked into play and now controlled my emotions. I wanted the sight to linger so that I could study and memorize it, and, yet, I was relieved when the eclipse ended and the familiar world returned.

James Fenimore Cooper, author of *The Last of the Mohicans* and other American classics, recalled seeing a total solar eclipse some twenty years after the fact. Even after those many years, he would write: "My recollections of the great event, and the incidents of the day, are as vivid as if they had occurred but yesterday."

It was 1806 and he was seventeen years old and living with his family at Lake Otsego in New York. The whole population was in a state of anxious expectation for weeks before the event. On the eve of it, he and his family could think or talk of little else. Every conversation quickly turned to a discussion of the movement of planets and the Moon and the coming eclipse.

That night, before the eclipse, Cooper looked at a star-filled sky and wondered how things would be different after the next day. And when the next morning came, family and neighbors gathered. Pieces of colored glass were handed out. And everyone turned their attention to the sky.

One person burst out with—as Cooper put it—"an exclamation of delight, almost triumphant" when the first thin piece of the Sun was seen to be missing. Then each person cried out in turn when they, too, saw it. Gradually, half, then three-quarters of the Sun was covered. When the Sun was almost gone, Cooper scanned the sky. He saw a spark glittering before him. "For a second I believed it to be an illusion, but in another instant I saw it plainly to be a star. One after another they came into view." It was noontime and the sky was filled with stars and the Sun had gone out.

Three minutes of darkness elapsed. "A breathless intensity of interest was felt by all." It was "a majestic spectacle" and "one of humiliation and awe." Then the stars retired and light returned. He likened "this sudden, joyous return of light, after the eclipse, to nothing of the

MASK OF THE SUN

kind that is familiarly known." It was not like the dawning of day or the end of a sudden storm. It was what one would "expect of the advent of a heavenly vision." He looked at his family and neighbors. He saw women standing with streaming eyes and clasped hands. The most educated men he knew stood silent in thought. Several minutes passed before anyone spoke, then it was in whispers. Cooper ended his account by stating, "Never have I beheld any spectacle which so plainly . . . taught the lesson of humility as a total eclipse of the sun."

A decidedly different account of a total solar eclipse came more than a century later from the British writer Virginia Woolf. The event was so widely publicized months before by the British press that more than three million people made the trek to north England to see it. Woolf, her husband, and three friends were among those who made the trip. They left by train at 10:00 P.M. on the eve of the eclipse—which would happen on June 29, 1927—leaving from Euston Railway Station at King's Cross in London. It was a five-and-a-half-hour ride to Richmond in North Yorkshire. It was impossible to sleep on the train. Woolf recalled in her diary that she passed the time smoking a cigar.

Along the way, as others would observe, people dispensed with the common greeting of "Hello" and replaced it with the single word, "Eclipse." At 3:30, the train reached Richmond near the center of the thirty-mile-wide eclipse path. Here a long line of motorcars and omnibuses waited, ready to take travelers to hilltops where there would be the best views of the eclipse. Woolf and her companions chose an omnibus. The morning was cold, and clouds covered the sky. They were taken to the moors, boggy and wet, high on Barden Fell several miles north of the town of Ilkley. Hundreds of others had already arrived. Some had obviously camped. None, as Woolf observed, were at their best, but still most seemed to feel a need to project a certain dignity.

They were strung out along the hilltop like statues. Most people stamped to keep warm. Others had brought bed sheets to keep them warm.

At 4:45 the Sun rose. "We began to get anxious," Woolf wrote. "We saw rays coming through the bottom of clouds." For a moment, she could see the Sun in full splendor and golden. But then clouds shifted and it was gone.

For the next hour, as she looked eastward, she could see that the Sun was in and out of clouds. To the west the sky was perfectly clear. Most people held watches in their hands. It was now almost time for the precious twenty-four seconds of totality to begin. The question was whether the Sun would make another appearance. Woolf felt she was being cheated.

It seemed that the Sun would make one more effort. For moment, she and others saw a thin bright crescent. Finally, the clouds won. The sky grew dark. Woolf was convinced this was the total eclipse. But was it? Then the sky grew darker still. She sensed that something more was about to happen. With the suddenness of a small boat capsizing and being unable to right itself, the light all around her went out. She now was at that mercy of the sky. "This was the end," she wrote. "The flesh and blood of the world was dead." But after the twenty-four seconds passed, with equal suddenness—Woolf recording that she felt "a great sense of relief"—light returned, at first somber, then with color.

On reflection, it had been an impressive adventure, picking oneself out of a London drawing room to ride a train for hours at night to stand and wait for dawn on one of the remotest moors in England. "Never was there a stranger purpose than that which brought [so many] together that June night."

She was back in London that evening. This time she had slept on the train.

The sense of excitement and anticipation that sent Woolf and millions of others into North England for a day to see a total solar eclipse can be traced back to Baily and Airy. Solar eclipses were in vogue after these two men made their trips in 1842. Never again would a solar eclipse pass over the Earth without someone traveling to see it, even if it meant a long journey to a distant part of the world.

CHAPTER EIGHT

A Simple Truth
of Nature

High on her speculative Tower
Stood Science waiting for the Hour
When Sol was destined to endure
That darkening of his radiant face
Which Superstition strove to chase,
Erewhile, with rites impure.

—William Wordsworth, after the partial
solar eclipse of September 7, 1820

Most historians give January 7, 1839, as the date when photography was invented, the day when François Arago, France's leading scientist and soon to be its prime minister, announced to the world that his fellow countryman, Louis Daguerre, had devised a chemical process that could imprint a permanent image onto a metal plate. I prefer to put the date when photography was invented a few months later on July 14 because it was on that day that the Parisian art critic Jules Janin told the world what it meant.

"No longer will anyone or anything be able to hide from a simple truth of nature," wrote Janin in the popular news journal *L'Artiste.* "And that is a little sad."

Sad or not, Daguerre's invention did transform the world. That August, he announced that he was giving his invention as a "gift" to the world. (He also received money from the French government the same month.) He then proceeded to describe how the process worked, that is, how one had to coat a highly polished copper plate with silver iodide, expose the plate to strong light for an hour or more, then hold the plate for several minutes over noxious fumes of mercury vapor to reveal an image. An account of the method and the materials needed was published soon after in newspapers across Europe and the United States, spreading news of the process around the world.

On September 20, a copy of London's *Literary Gazette* that described Daguerre's method arrived in New York. The next day John Draper, a professor of chemistry at New York University, was at the Astor House where he saw a copy of the *Literary Gazette* and read of Daguerre's method. The day after that Draper purchased the necessary chemicals and metal plates and had built a camera—the first camera ever in the United States—from a cigar box, using a glass lens from a pair of eyeglasses. After one more day, on September 23, 1839, Draper was taking his first photographs.

He began by taking portraits of a laboratory assistant who had to sit motionless for half an hour in strong sunlight. Draper improved the process, making the chemicals more sensitive to light, and he built cameras with larger lenses. By the next spring he was taking portraits of family members. The sitting time had been reduced to ninety seconds. He was now convinced he could take photographs at night. The first subject he chose was the Moon.

For several clear nights in early 1840 he climbed to the roof of the main university building. After many trials he finally succeeded in getting an image of our nearest celestial neighbor on a small piece of copper plate. The exposure time was twenty minutes. The Moon's image is an inch in diameter. It is bright with a few irregular dark spots. Draper exhibited it in April 1840 at New York's Lyceum of Natural History. People were astounded by what he had accomplished.

Of course Draper was not the only person improving on Daguerre's process—and pointing cameras to sky. Though the first full image of the Sun would not be made until 1845—the problem was building a camera with a shutter speed that was fast enough to significantly reduce the Sun's intense light—and the first star would not be recorded until 1850, the first image of a partially eclipsed Sun was made in 1842 by Alessandro Majocchi in Milan. During that eclipse, Majocchi exposed a plate just a few minutes before totality. The result showed a thin crescent of sunlight. Majocchi also tried to record the faint light of the corona, but totality lasted only two minutes and not the slightest trace of the corona can be seen on the exposed plate.

The sensitivity of Daguerre's process continued to improve. In 1851 Johann Berkowski, using a small telescope at the Königsberg Observatory in East Prussia, took the first photograph of a total solar eclipse. The exposure time was eighty-four seconds. The image of the eclipsed Sun is only one-third of an inch in diameter, and, yet, when studied through a magnifying glass, one can see three or more solar prominences extending slightly beyond the Moon's limb. There is also the faintest hint of the white corona.

Now there was interest in making a special effort to photograph total solar eclipses. And that interest would eventually grow into a stampede.

It was the manufacturing of playing cards that led Warren De La Rue into astronomy. In 1831 his father, Thomas, received a patent for a process that applied a glaze that made playing cards more durable and less susceptible to marks produced when handling the cards. The new cards proved popular. And the De La Rue Company was soon producing more than a hundred thousand decks of cards each year. Needless to say, it made the De La Rue family wealthy. To keep it that way, Thomas De La Rue sent his son Warren on trips to promote the playing cards to potential distributors and to seek out improved ways to manufacture the cards.

On one such trip, in 1840, the younger De La Rue was in the workshop of the well-known Scottish inventor James Nasmyth. Nasmyth had already invented a number of devices popular in manufacturing, such as the steam hammer that was used to shape large iron shafts in shipbuilding. He was also an expert in the application of paints and resins to metal, wood, or paper. It was the last that interested De La Rue. Nasmyth had recently perfected a way to apply a thin transparent coat of white lead to paper, and De La Rue saw its obvious application to the manufacturing of playing cards. But while he was visiting Nasmyth he noticed something else. The Scottish inventor was then casting mirrors to be used in telescopes. Nasmyth gave him a demonstration, showing how one of his mirrors could be used to project an image of the Sun on a wall. De La Rue, still unlearned in astronomy, inspected the image and pointed out several dark blotches suggesting they were imperfections in the mirror. Nasmyth corrected him. Those dark "blotches" were sunspots. Nasmyth was showing him the true surface of the Sun. De La Rue immediately asked Nasmyth to cast a mirror for him.

With the help of Nasmyth and several astronomers from the Royal Society in London, De Le Rue designed and built his own telescope and enclosed it inside his own observatory building. He became a regular observer of the night sky, studying star clusters and making detailed drawings of planets, especially, Saturn. He also photographed the Moon, using a new process known as wet collodion that was more sensitive to light than Daguerre's process.

While pursuing his interest in astronomy, De La Rue also continued to travel. In 1858 he was in Prussia, again on a business trip, and took time to visit the Königsberg Observatory where he was shown the image of the total solar eclipse made by Berkowski. The size of the Moon's image was much smaller than what he was taking with his telescope. And it was clear that Daguerre's process gave much less detail than the wet-collodion process that he was using. He would see the prominences, but just barely. He asked what they were. The astronomers said that they might be clouds hanging in a tenuous lunar atmosphere or distortions caused by the Earth's unsteady atmosphere or, perhaps, something huge rising high above the Sun's surface. The astronomers at Königsberg also showed him a pamphlet that gave details about the next total solar eclipse. It would occur in two years on July 18, 1860, when the Moon's shadow would pass over the Atlantic Ocean and northern Spain, then continue across the Mediterranean Sea and Africa. De La Rue realized this was an opportunity to determine exactly what the prominences were. But that would require transporting tons of equipment, both a telescope and a complete darkroom, to a location somewhere along the path of the Moon's shadow so that a detailed photograph of the total eclipse could be taken.

Fortunately, the path across northern Spain coincided with that of a newly completed railway line, which would ease the burden of transporting the equipment. De La Rue also discovered there was interest in England for such an adventure. Astronomer Royal George Airy was eager to accompany De La Rue. Airy used his considerable influence at the British Admiralty to find money to fund the project. He also persuaded the Admiralty to assign a ship to carry De La Rue, a cadre of assistants, and other interested people and the equipment to Spain.

They sailed from Plymouth Sound on July 7, eleven days before the eclipse. Two days later they arrived at Bilbao on the northern coast of Spain. Onboard were thirty large wooden boxes filled with equipment. The customs official in Spain lifted the lid of one box, inspected the inside, then let the boxes and De La Rue and the others enter the country.

For his observing station, he chose the village of Rivabellosa about thirty miles south of Bilbao and on the southern side of the Pyrenees, in order to avoid the clouds that often build up against the northern slopes of the mountains. He had a portable structure erected, one of his own design. It had a canvas cover that would be wetted to lower the temperature inside. One half of the structure, where the canvas could be pulled back, housed the telescope, itself a massive instrument. The other contained an entire darkroom with sinks, bottles of chemicals on shelves, and a cistern that held sixteen gallons of distilled water. The wet-collodion process required that photographic plates be prepared immediately before exposure and developed immediately after.

As De La Rue knew, it was selecting the wet-collodion process, instead of the more reliable Daguerre process, that was the most controversial part of the trip. It required the use of a wider range of chemicals, some potentially explosive, that had to be mixed in a highly precise manner to form a syrupy solution, the collodion. The solution was then carefully poured onto one side of a clear glass plate to prevent smudges or bubbles, and the plate dipped into a bath of silver iodide. After the bath the prepared plate was put in a dark sleeve that was slid into the camera. The plate was then exposed and removed and put into a bath of ferrous sulfate to wash away whatever silver iodide remained, then dipped in a bath of potassium cyanide to fix the image, then washed in distilled water, dried over candles, and, finally, covered with lacquer. It was a complicated process and any mistake could distort the image or cause streaks to appear. And yet it was far more sensitive to light than the Daguerre process and could record much finer detail.

And then there was the question of the exposure time. De La Rue had no guidance as to how sensitive the photographic plates would be. Would half a minute be enough? If he exposed a single plate for three minutes, then there would be only one photograph of totality. In the end, he decided on a middle ground. He would record two plates for one minute each.

There was also a worry whether the equipment would work. For days before the eclipse, De La Rue rehearsed with his assistants what their exact tasks would be during the few minutes of totality. Others, who would later also travel great distances to photograph eclipses, would tell how they suffered from insomnia for many nights before the event, worried about equipment and other matters. Some reported having nightmares. One astronomer who traveled from England to Thailand to photograph an eclipse recalled that things went so badly that he "could not help sitting down and having a good cry."

And then there was always a concern about the weather. Two days before the eclipse, De La Rue saw one of the "most awful thunderstorms" he had ever witnessed. The following day was completely overcast. The morning of the eclipse it seemed hopeless that the weather would clear. He watched the sky with anxiety. "I am free to confess," he would write, "that *my* nerves were in the most feverish state of agitation." Not until noon was there any indication of clearing, then it was gradual. Then, just moments before the Moon began to creep over the Sun, in the fashion of a miracle, the clouds dissolved away and the sky was perfectly clear. De La Rue and his assistants proceeded to do their tasks.

Twenty photographs were taken of the partially eclipsed Sun. Then, seconds before totality, the twenty-first glass plate with collodion was prepared. The assistant inside the darkroom placed the plate inside a dark sleeve and handed it to a second assistant. He carried it the few feet to a third assistant standing partway up a ladder and he handed it to De La Rue, who was positioned to insert it into a holder on the telescope. A fourth assistant then opened the shutter that was at the opposite end of the telescope. A fifth assistant who was watching a chronometer then began to call out the seconds, timing the exposure. Meanwhile, with a minute to relax, De La Rue looked through a second, smaller telescope and studied the limb of the Moon.

As the precious seconds went by, he could see several bright red prominences. As he continued to watch, he could see one side of the dark limb of the Moon covering up a small prominence while, on the other side, a new small prominence was coming into view: A clear sign that the prominences were attached to the Sun, not the Moon. But would he get the photographic evidence to prove it to others?

At the end of a minute, De La Rue removed the plate, handed it back to one of the assistants so that it could be developed immediately, then received a new plate and inserted it into the holder on the telescope. Another minute was passing. Midway through it, De La Rue could hear the assistant inside the darkroom, a Mr. Reynolds, shout out that they had an image on the first plate. As De La Rue later said, those words filled him "with a thrill of intense pleasure." Then the unthinkable happened. De La Rue was politic enough never to say who was responsible, but someone bumped the telescope. The second plate, which was needed to compare with the first, would have a double image, and so it was ruined. After months of preparation and the expenditure of so much money and effort, he had fallen short. He would have only his visual observations to report about the solar origin of the prominences and that alone would not be regarded as conclusive.

But there is a happy ending to this story. Fortunately, another astronomer, Father Angelo Secchi of the Roman College at the Vatican, had gone through the same effort of planning and carrying out an expedition to photograph the 1860 solar eclipse. He had stationed himself 250 miles away from De La Rue on the eastern coast of Spain and he also was able to obtain a single photograph of totality. Though his photograph was not as good in quality as the one taken by De La Rue, years later, De La Rue took his glass plate to Rome and compared the two. Without a doubt, the comparison showed that the Moon had shifted with respect to the pattern of prominences.

———•———

Today it is easy to trivialize what De La Rue and Secchi had done. But one must remember how little was known about the physical character of the Sun and the Moon when they were doing their work.

In 1836 Baily had taken precious moments during totality to carefully study the dark silhouette of the Moon through his telescope, wondering if he might see a few bright spots of sunlight that would indicate the Moon's interior was cavernous. There were also those who thought the Moon's surface consisted of a thick layer of material that was so loose that if someone tried to stand on it, they would immediately sink down and disappear.

Ideas about the Sun were even stranger, that is, by our standards today. In the 1810s William Herschel, the discoverer of the planet Uranus and then one of the world's foremost astronomers, had studied sunspots and concluded that they were openings to a cold dark interior and that the part of the Sun that we could see was actually a thin shell that consisted of a vast luminous ocean. By the 1860s the idea of a cold interior had been discarded in favor of a luminous ball, but exactly what was the Sun? Was it a solid, a liquid, or a gas? The blinding brilliance of sunlight had prevented progress on this question. And that is why a study of total solar eclipses took center stage. One should be able to learn much about the Sun by studying the prominences. And there was a new tool, developed in the 1850s, that would lead the way—spectroscopy.

As in many things, Newton had led the way when he made a small circular hole that allowed a beam of sunlight to enter a darkened room, then using a prism spread the sunlight into the familiar rainbow of colors. In 1814 Joseph von Fraunhofer, a glassmaker in Germany, greatly improved on Newton's original experiment by, among other things, using a thin slit instead of a small circle. He also used multiple glass prisms that allowed him to make wider rainbows that he then studied in detail, noticing that hundreds of dark lines appeared. The more prominent of these dark lines he designated with capital letters from *A* to *K*, omitting *I* and *J* because of the difficulty in distinguishing the two letters in print.

The mystery of the dark lines was solved in 1859 by Gustav Kirch-hoff. He showed that a unique set of dark lines could be associated with each chemical element. For example, the C and F lines identified by Fraunhofer corresponded to strong lines produced by Kirchhoff when he heated hydrogen gas in a glass tube. The E line Kirchhoff associated with iron, the G line with calcium and the D line with sodium, such as that found in table salt. To make these determinations, Kirchhoff had invented a new type of scientific instrument, the spectroscope. Suddenly, the world of inquiry became much larger for astronomers. They could now use spectroscopes attached to their telescopes and reveal what some people had thought would never be possible: to determine the composition of stars.

It was also obvious that his new tool could be used to determine the nature of solar prominences. And so ten different expeditions were organized by four different countries to send astronomers to photograph and make spectroscopic measurements of the next total solar eclipse, one that would occur on August 18, 1868.

The path of totality in 1868 would not be through Europe as it had been in 1851 and 1860. Instead, it would begin at the southern end of the Red Sea at the city of Aden, cross the Arabian Sea to southern India, then cross the Bay of Bengal to Thailand, and finally to Borneo and the islands of the New Hebrides. German astronomers, after studying a map of the eclipse path and considering possible weather conditions, decided to send two expeditions, one to Aden and the other to India. The Austrians sent one expedition, also to Aden. The British, in political control of India since 1858, sent four expeditions to the subcontinent. The French were divided internally. The Bureau of Longitude in Paris sent Pierre Janssen and a few others to India—much more about him later—while the Paris Observatory sent the main French contingent of more than a dozen astronomers to Thailand at the invitation of King Mongkut.

Mongkut was the fourth monarch of Thailand, his rule beginning in 1851. It was during his reign that Western ideas began to influence Thailand. Mongkut attempted to bridge the two worlds, maintaining ancient Buddhist traditions while finding ways to democratize the country. He is the king depicted in the popular musical *The King and I*. The British woman who educated his children was Anna Leonowens.

When he learned of the eclipse some years in advance, Mongkut was determined to make Thailand one of the focal points where scientists would come and view the event. He spent many months computing exactly where in Thailand the eclipse would be seen. Then he chose a spot along the center of the path, Waghor village on the Malay Peninsula, where he would lead his royal court to view the eclipse. He invited scientists and dignitaries from every major European country to attend. Only the French accepted the invitation along with the British Governor of Singapore.*

A large area had been cleared in the jungle where Mongkut, his court, the French scientists, and the British Governor would view the eclipse. The scientists erected their telescopes. The others partook of excellent dishes prepared by a French chef who resided in Bangkok. Their drinks were cooled with ice, also brought from Bangkok.

The day of the eclipse began cloudy, but cleared in time for the eclipse. Immediately after, Mongkut left offerings for those local deities that had provided the clear skies. He then returned to Bangkok where he soon became ill. In *The King and I* the king died from some unknown cause. In reality, Mongkut died from malaria contracted during the trip to see the eclipse. But his effort is remembered today. In Thailand, the day of the eclipse, August 18, is National Science Day, in remembrance of King Mongkut and his prediction of the 1868 total solar eclipse.

The French scientists who joined Mongkut did get what they wanted. In particular, Georges Rayet, who the previous year had

---

* Anna Leonowens was in the United States at the time of the eclipse.

identified a new type of star, known today as a Wolf-Rayet star,[*] aligned the slit of the spectroscope to a solar prominence and saw many strong lines. And he was not alone. Astronomers at other locations who were using spectroscopes also saw a set of strong lines in the spectrum of solar prominences. Using arguments that Kirchhoff had constructed, each one of them immediately knew that, based on the solar prominence he was looking at, as well as the behavior of its outer surface, the Sun must be a gas. Furthermore, the strongest line in the spectra from the solar prominences was a red line due to hydrogen, which was the reason the prominences seen during eclipses were a bright red. At that point Rayet packed up his equipment and returned home. And so did all of the other observers of the 1868 eclipse who were scattered across the world, that is, all of them except one—Janssen in India.

During the eclipse he was struck by the strength of the lines he was seeing in the spectrum of the solar prominences, and the thought occurred to him immediately that it might be possible to see them *without* an eclipse. Unfortunately, the weather clouded after the eclipse, preventing him from attempting anything more that day.

He rose at three o'clock the next morning excited by the prospect of what he would soon do. He prepared his telescope and spectroscope. He watched the Sun rise. As soon as it was free from the mists near the horizon, he positioned the telescope so that it pointed close to the Sun's limb. He looked thought the spectroscope. *Voila!* There they were, a strong set of lines in the spectrum just as he had seen them the previous day during the eclipse. He had revolutionized how the Sun could be studied.

He remained in India another month, perfecting the technique and observing the spectrum of solar prominences whenever weather permitted. On September 19, he wrote about his discovery in a letter to Jean-Baptiste Dumas, the secretary of the *Académie*

---

[*] These are now known to be highly luminous, massive stars that derive their energy from the fusion of helium instead of hydrogen.

des sciences in Paris. The letter took five weeks to reach Paris from India, arriving on October 26. By great coincidence, on the same day, at the moment the letter arrived, Warren De La Rue was reading a letter at a meeting of the *Académie des sciences* that he had just received that same day from his fellow countryman Norman Lockyer. In the letter, Lockyer, who had been too ill to make the trip to India to observe the 1868 eclipse, stated that he had just discovered how to use a spectroscope to observe solar prominences *without* an eclipse.

Consider the animosity that had often existed between the two countries, including between their scientists—whether the discovery of the planet Neptune in 1846 was owed more to the work of the British astronomer John Adams or the French astronomer Urbain Le Verrier being a recent one. Yet after the two letters were read at the Academy, one from Janssen and one from Lockyer, members decided that both men deserved credit. In 1872 a medal was commissioned to commemorate the joint discovery and a close friendship was also cemented between the two men.

———

Lockyer now took the study of the spectrum of solar prominences one step further. The three strongest lines were always the same. The one at the red end of the spectrum Lockyer was able to identify as the C-line in Kirchhoff's earlier work, which meant it was due to the presence of hydrogen. An equally strong line in the blue was Kirchhoff's F-line and it too was due to hydrogen. But the third line, this one in the yellow, seemed at first glance to correspond to Kirchhoff's D-line, which was due to sodium. But as Lockyer improved his observations, making better measurements of exactly where the lines were along the rainbow spectrum, it became clear the yellow line in the spectrum of solar prominences was shifted a small amount and did not correspond exactly with the D-line of sodium. What might it be?

Lockyer sought the help of Edward Frankland, a chemist at the Imperial College of London. It was hoped together they would be able to find the meaning of the mysterious yellow line. They began by varying the pressure of hydrogen and measuring the spectrum. No yellow line appeared. They thought the motion of gases on the Sun might account for it. But that, too, failed to explain the yellow line. Finally, Lockyer concluded the yellow line might be due to a new chemical element, one that might only exist on the Sun. And so he named it for the Greek god of the Sun, Helios, that is, he named the new element *helium*.

And that is where the matter stood for the time being. Lockyer was often criticized by his contemporaries for resorting to the extreme conclusion that some chemical elements that exist in the universe might not exist on Earth. The idea actually floundered in scientific circles. When it was mentioned at all, scientists often referred to it as "the hypothetical helium." Dmitri Mendeleev, the Russian chemist who was revolutionizing chemistry, gave one of the harshest critiques when he said there was no place for Lockyer's helium on the periodic table of chemical elements he was then developing. But in 1895 this changed.

William Ramsay of University College London was assaying a sample of pitchblende, a dark rock that is often associated with silver. He began by crushing a portion of the rock into a powder and then treating the sample with sulfuric acid. Gas started to bubble up from the sample. At first he thought it must be carbon dioxide, a gas commonly released when acid is poured onto a rock sample. But the rate of bubbling was slow. He collected the gas and examined it with a spectroscope, discovering that it showed a strong yellow line—the same yellow line Lockyer had seen in the spectrum of solar prominences.

In 1900, now with overwhelming evidence, Mendeleev finally included helium on the periodic table of chemical elements.

———•———

Both Lockyer and Janssen continued to pursue total solar eclipses into the next decade. One would occur just before Christmas 1870 and be visible from southern Spain, Sicily, and part of North Africa. Lockyer organized four British expeditions to be sent to Spain, Gibraltar, Algeria, and Sicily. Lockyer was in charge of the Sicilian party. He sent a message to Paris asking Janssen if he might be able to join the expedition in Algeria. The Franco-Prussian War had just started. And Paris was under siege.

Working through the English Foreign Office, Lockyer managed to convince the Prussian Prime Minister Otto von Bismarck to grant Janssen safe conduct out of the city. Janssen declined the offer. Instead, he chose a different way out of Paris—by balloon.

Balloons had been used during the siege of Paris to fly military and government officials out of the city. On October 7 the minister of war, Léon Gambetta, flew out of Paris to organize an offensive against the Prussians from Tours. On December 2 Janssen would do the same.

Paris had long since run out of silk balloons. More than a hundred seamstresses now worked to sew together bolts of cotton cloth to make giant air bags that were varnished with linseed oil to make them impermeable to the coal gas they would contain. To transport people and equipment, a wicker basket, four by four feet in area and three feet high, hung beneath the air bag. Sailors from the French navy were used as pilots because flying through the air by balloon was considered to be comparable to sailing the seas. The balloon Janssen would fly in was named *Volta*. The name of his pilot is recorded in history simply as Chapelain. The only baggage was the three hundred pounds of astronomical equipment Janssen would need packed tightly inside four wooden cases. Tightly wadded, shredded paper was wrapped around each piece of equipment to protect it from sudden jarring.

Janssen and Chapelain left Paris at 6:00 o'clock on December 2. They flew over the Prussian lines at about 2,000 feet, still in the dark early part of morning twilight. At 7:35 the Sun rose. The wind was strong that day. In just five hours, Janssen and Chapelain had traveled more than two hundred miles. They decided to land near the village of Briche-Blanc, Chapelain releasing gas from the gasbag.

Local people surrounded them immediately asking if they had just come from Paris and what conditions were like in the city.

A special train was arranged to take Janssen and his equipment to Nantes and from there to Tours where the French government was now located. In Tours he met with government officials, including Gambetta, giving them verbal messages he had memorized from officials still in Paris. It was probably because he knew the message that he had been unwilling to pass by land through Prussian lines.

At Tours another special train was arranged that took him to Bordeaux, then to Marseille where he boarded a ship and sailed for Oran, Algeria, arriving there a few days before the eclipse. He set up his equipment. He practiced pointing the telescope, using the spectroscope, and taking photographs of the Sun. The day of the eclipse, December 22, 1870, it rained all day.

Lockyer did not see the eclipse from Sicily either. The ship he sailed on, HMS *Psyche*, struck a submerged rock in Naples Bay in Italy. Fortunately no lives were lost and the astronomical equipment was saved, but the ship was wrecked. Lockyer did find a way to reach Sicily overland, but on the day of the eclipse the sky there was also cloudy. Among the few people who saw the total solar eclipse of 1870 were those who had remained in Naples and who were trying to salvage the wrecked *Psyche*. They had a perfect view.

The U.S. Navy dirigible *Los Angeles*. 1929. *Courtesy U.S. Navy.*

Several of the crew members who flew on the *Los Angeles*
during the 1925 total solar eclipse. Chief Quartermaster Peterson
is second from right. January 24, 1925.
National Archives and Records Administration Ref. No. 80-G-460206.

Peterson demonstrating the movie camera he used to film the solar eclipse. *National Archives and Records Administration Ref. No. 80-G-460210.*

Peterson on top of the *Los Angeles* with the movie camera.
National Archives and Records Administration Ref. No. 80-G-460202.

The nine planets (*navagrahas*) carved on the lintel inside the sanctum doorway of Mukteshvara Temple, Bhubaneswar, India. Beginning from the left, the first seven images represent the Sun, the Moon, Mars, Mercury, Jupiter, Venus, and Saturn. The invisible planets of Rahu and Ketu are depicted at the far right.
Courtesy Alamy Stock Photo.

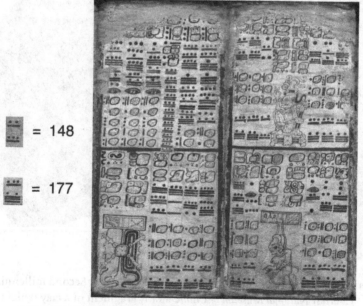

= 148

= 177

Pages 52 (left) and 53 (right) from the Dresden Codex.
The number 148 is shown once at the bottom of page 53 and
the number 177 eight times at the bottom of both pages.

The three largest fragments of the Antikythera Mechanism. The largest gear on the center fragment has 223 gears, the same number as the number of full moons in a Saros cycle. *Courtesy Cardiff University/Alamy Stock Photo.*

Text B (BM 34757)

LEFT: Oracle bone (ox scapula) from the Shang Dynasty. Second millennium B.C.E. *Courtesy National Museum of China.* RIGHT: Fragment of a clay tablet from Mesopotamia. First millennium B.C.E. *Courtesy Trustees of the British Museum.*

Map prepared by Edmond Halley that shows the predicted path of the Moon's shadow across England in 1715. *Courtesy Royal Astronomical Society of London.*

Eclipse expedition to Denver, Colorado, led by Maria Mitchell of Vassar College. July 29, 1878. *Courtesy of the Nantucket Maria Mitchell Association.*

Eclipse expedition to Jeur, India, led by William Campbell of Lick Observatory. January 22, 1898. *Courtesy Special Collections, University Library, University of California Santa Cruz, Lick Observatory.*

Drawing of the solar corona
by José Joaquín de Ferrer at
Kinderhook, New York.
June 16, 1806.

Drawing of the
solar corona by
Samuel Langley at
Pikes Peak, Colorado.
July 29, 1878.

Composite drawing of the solar corona by Mabel Loomis Todd,
based on 100 sketches made by other observers. January 1, 1889.

True corona during a total solar eclipse. July 11, 1991. *Photo by John Dvorak.*

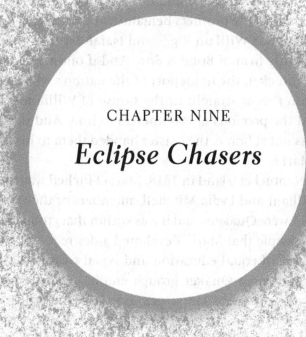

CHAPTER NINE

Eclipse Chasers

It sounded as if the Streets were running
And then—the Streets stood still—
Eclipse—was all we could see at the Window
And Awe—was all we could feel

— Emily Dickinson, 1877

I n the first half of the nineteenth century, as soon as a ship
reached a major port, the ship's master gathered the collection of chronometers on board and took them ashore
to a nautical chandler to be repaired and rated. Rating meant
determining the daily rate a chronometer was gaining or losing

compared to a master clock on shore, the master clock reset frequently by the chandler making astronomical observations throughout the day and the night. In New York, the best raters of chronometers were the brothers Benjamin and Samuel Demilt. In Philadelphia it was William Riggs and Isaiah Lukens. In Boston it was the family firm of Bond & Son. And if one arrived on the island of Nantucket, the homeport of the nation's whaling fleet, a ship's master went straight to the house of William Mitchell and handed the precious chronometers to him. And if William Mitchell was not at home, the master handed them to his teenage daughter Maria.

Born on Nantucket Island in 1818, Maria Mitchell was the third child of William and Lydia Mitchell, members of the Society of Friends. They were Quakers, and it was within that group of highly disciplined people that Maria developed a desire for learning in an atmosphere of equal education and equal treatment of both sexes. Whereas most Quaker groups emphasized the learning of biblical messages and languages, on an island where many were destined to sail to remote parts of the oceans, Nantucket Quakers were taught the rudiments of navigation and its attendant science—astronomy.

Mitchell learned to read a compass almost as soon as she learned how to read a book. As a child, she understood how a clock worked and knew how an understanding of time came from movements of the Sun and the Moon in the heavens. Some of her earliest written notes are found in her copies of Bridge's *Treatise on Conic Sections*, Hutton's *Mathematics*, and Bowditch's *American Practical Navigator*. At age seventeen she became the librarian on Nantucket, a job that included ample free time to read celestial mechanics and astronomy. In 1847, using her father's highly treasured telescope, made by John Dollond of London, one of the world's premier manufacturers of telescopes, she discovered a comet. It was the first time a comet had ever been discovered using a telescope. For that accomplishment and for other astronomical observations made with the Dollond telescope, in 1848 Mitchell was elected an honorary member of the American Academy of Arts and Sciences.*

In 1865 Matthew Vassar, a wealthy brewer, was planning to start an all-female college in his hometown of Poughkeepsie, New York. One of the first people he approached to teach at his new college was Mitchell, but when she seemed to be too assertive Vassar had second thoughts. He sent a friend, a Mr. Rufus Babcock, to interview her and her father to decide what type of person she was. After an evening with both Mitchells, Babcock was pleased to report that "Miss Maria is not such a poor miserable 'blue-stocking' as to know nothing else but astronomy." She had prepared dinner that night and "presided in all the housewifery of [the Mitchell's] cozy establishment without parade and without any apparent deficiencies." And so she was hired as the first faculty member of Vassar College.

Matthew Vassar had a large telescope built for her, and she lived inside the observatory building. During her tenure at Vassar College, she educated a generation of women in the sciences, including mathematician and logician Christine Ladd and MIT chemist Ellen Richards. She insisted on learning by observation, not by rote. She gave no grades. And refused to take attendance.

Given her inclination toward all things astronomical, Mitchell had the expected interest in observing eclipses. Her first experience came at age twelve. On February 12, 1831, an annular eclipse passed directly over Nantucket. The day was icy but clear. Her father prepared by removing a window from an upstairs room in the house and setting up the Dollond. Maria sat next to her father and counted out the seconds. From the timing of the eclipse, they could determine the longitude of their house on Nantucket where ship chronometers were rated.

---

* No woman would be elected to the American Academy of Arts and Sciences until 1943 when four women were added as members: astronomer Cecilia Payne-Gaposchkin, psychologist Augusta Fox Bronner, Radcliffe College president Ada Louise Comstock, and writer Willa Cather. Anthropologist Margaret Mead was elected two years later in 1945.

Mitchell's next chance to measure a total solar eclipse came years later in 1869, after her appointment to the faculty of Vassar College. She gathered seven students and they headed west by train to Burlington, Iowa.

The head of the scientific effort at Burlington was John Coffin, superintendent of the Nautical Almanac Office in Washington, D.C., the bureau of the government that published annual tables of the positions of the Sun, the Moon, the planets, and several bright stars, as well as details about eclipses. According to the almanac, on August 7, 1869, the path of a total solar eclipse would pass over Alaska and western Canada then sweep across the continent from Montana to North Carolina. In Iowa, where good weather was expected, the period of totality would last nearly three minutes.

Congress had appropriated $5,000 for eclipse observations. Astronomers from Dartmouth College in Hanover, New Hampshire, and Lehigh University in Bethlehem, Pennsylvania, joined Coffin in Burlington. They arrived on the same train as Mitchell and her students. The men, who would be taking photographs and making spectroscopic measurements, set up their equipment at the corner of Seventh and Elm Streets. Today a stone monument with a bronze plaque commemorates their efforts. For an unexplained reason, Mitchell and her students set up their equipment—three chronometers to time the eclipse and two telescopes to visually observe it—a few miles northwest of Burlington on the grounds of the Burlington Collegiate Institute. Coffin's daughter, Louisa, joined them.

Mitchell positioned one of her students on the roof of the college to watch for the approach of the Moon's shadow and to make general observations of changes in the landscape. Another was put on the roof of a nearby private house for the same reason. Louisa Coffin took temperature measurements. The other six paired off; one would look through a small telescope—Mitchell used the Dollond—while the other counted out half-seconds.

Seconds before the eclipse, Venus suddenly appeared near the Sun. Mitchell could hear cattle begin to low. She could see fireflies in the foliage. Instantly the corona burst forth, encircling the darkened Sun with a soft white light and sending off streamers. She recorded the location of prominences. One was shaped like a spiral, a whorl that resembled a half-open morning glory. One of the students listened as Mitchell called out her observations and took notes. After the eclipse, the observations were compiled. The approaching shadow had given the landscape the general appearance as before a thunderstorm. It was dark enough that almost everyone had seen fireflies. A Miss Reynolds had heard crickets chirping. Detailed descriptions had been recorded of the prominences and of the corona. Those who were watching the chronometers determined that the total phase of the eclipse had started twenty-three seconds earlier and had lasted thirty seconds longer than predicted.

Mitchell wrote a summary of the work for *Friends Intelligencer*, a monthly Quaker magazine. She began by giving a primer about eclipses and why and how often they occurred. She recalled how on the day of the 1869 eclipse the barometer was rising, an indication of clear weather. And how Miss Coffin had measured a temperature drop of five degrees Fahrenheit (about three degrees Celsius). And how, according to Miss Blatchley who was atop the college building, the nearby Mississippi River had turned a leaden hue and birds had flown to a bell tower and fluttered about. But most importantly, Mitchell commented at the end of her summary about the ability to do one's assigned work while surrounded by so many unusual phenomena during a solar eclipse.

She quoted the Scottish astronomer Charles Smyth who had witnessed the 1851 total eclipse from Norway and who had said how the sight was so overpowering that a man might forget an assigned task and spend those previous minutes looking around. But Mitchell assured her readers that her "party of young students" would not have turned to look, even if the earth had quaked. And why was that, Mitchell asked.

She did not hesitate in giving the answer: "Because they were women."

———•———

The astronomers who had gathered at Burlington, Iowa, were by no means the only ones who were looking up and timing and taking photographs and studying the spectrum of the 1869 solar eclipse. Other groups could be found in Des Moines, Ottumwa, and Mount Pleasant, all in Iowa, and in Springfield, Illinois, and Louisville and Shelbyville, in Kentucky. Cleveland Abbe, director of the Cincinnati Observatory, had traveled to Sioux Falls, then called Fort Dakota, in the Dakota Territory, a town of abandoned soldiers' barracks and a half-dozen occupied houses, to observe the eclipse. The few residents of Fort Dakota gathering around him as he watched for Baily beads, prominences, and the corona.

The person who traveled the farthest to view the eclipse was Asaph Hall of the Naval Observatory in Washington, D.C. The eclipse track crossed just south of the Bering Strait between Siberia and Alaska, the latter purchased just two years earlier from Russia by the United States. The terms of the sale gave the latitude of the Bering Strait and described several geographical points, but it made no mention of the longitude because no such determination had yet been made in this remote part of the world. Hall's goal was to use the 1869 solar eclipse to change that.

He left New York by steamship on May 21, cross the Isthmus of Panama by train, then sailed again by steamship to San Francisco, arriving on June 12. Then it was nineteen days of preparations, much of it spent rating the ten chronometers he would carry with him. On June 19 he left San Francisco on the USS *Mohican*, a Navy warship that carried a complement of 160 officers and sailors. As the destination, Hall had chosen Plover Bay (today Providence Bay) in Siberia because it was close to the central path of the eclipse and because it was one of the few good harbors in the North Pacific. The sea passage took thirty-one days. The ship arrived on July 30. After a day of inspecting several potential observing sites, Hall chose a sand spit where he had the ship's carpenter construct a small shelter.

On the day of the eclipse—it was August 8 in Siberia—the morning began clear, but clouds soon appeared, and, by the time the Moon started to move in front of the Sun, the entire sky was covered. It remained that way until an hour after the Moon had moved off the Sun when the sky cleared and was cloudless until sundown. The *Mohican* left that night at midnight, arriving back in San Francisco on September 21. Hall had been at sea for 102 days. As his effort would be summarized in the official report, "the observation of the eclipse was not so perfect as we desired."

———·———

If one were to characterize the response to the 1869 solar eclipse as a gala affair, then the next total solar eclipse across the United States, which occurred just nine years later on July 29, 1878, was an unabashed celebration. About one hundred astronomers took part, positioning themselves at twelve different stations.

The eclipse path crossed the middle of the United States once again, though this time it swept along a north-south path over the eastern ranges of the Rocky Mountains from Yellowstone in Wyoming to central Colorado where it ran across Pikes Peak and the towns of Boulder, Denver, and Colorado Springs, then over the dry grassy plains of the Oklahoma Indian Territory and east Texas before entering the Gulf of Mexico between Galveston and New Orleans. Railroad companies offered tickets at half price to those who could prove themselves to be astronomers and were on their way to view the eclipse. Newspaper editors contributed to the promotion by reminding readers that there were obvious advantages to seeing an eclipse through the clear air of the high altitudes and that, at least in Colorado, each astronomer would have "a peak to himself."

And they came, these "wise men from the east," as local newspapers called them. Almost the entire staff of the United States Naval

Observatory went west, staff members assigned to one of eight observing stations. Asaph Hall of the Naval Observatory went to La Junta, Colorado. Edward Holden, also from the Naval Observatory, went to the mining town of Central City, Colorado, and received permission to set up his equipment on the flat roof of the three-story, 150-room Teller House Hotel. He was joined there by Charles Hastings, a professor of physics at Johns Hopkins University, and Edgar Bass, a professor of mathematics of the United States Military Academy at West Point. The evening after the eclipse, they invited the citizens of Central City up to the roof and showed them the planet Jupiter and several star clusters.

Professor Charles Young and several graduating seniors from Princeton University went to Cherry Creek, Colorado. Simon Newcomb, the superintendent of the Nautical Almanac Office stationed himself at Separation, Wyoming. David Todd, professor of astronomy at Amherst College in Massachusetts, went to Dallas, Texas. Others would find themselves near the towns of Creston, Wyoming, Las Animas, Colorado, and Melissa, Tyler, and Willis in Texas.

Samuel Langley of the Allegheny Observatory in Pennsylvania and several others chose Pikes Peak. Over a period of five days, they made the repeated eighteen-mile hike to the 14,000-foot summit carrying observing equipment, tents, bedding, and food. The summit itself is a plain, a few acres in extent, covered with sharp rocks and fractured boulders, and so little could be done to protect themselves from the constant strong wind. Added to the misery, it snowed, hailed or rained during most of the first several days they occupied the summit. The weather was so bad that Langley used lard to cover the iron parts of the equipment, then wrapped them in canvas to keep them in working order. He and others also suffered from altitude sickness, as he wrote it, "with a great difficulty of breathing and greatly increased action of the heart." They felt constant and severe headaches and "nearly every symptom that attends sea-sickness."

The weather of course was a concern for everyone, and a subject of constant conversation, no matter where they were. An editor at the *Denver Tribune* searched the archives of his newspaper and determined, going back six years, that the date of July 29 had been clear on four occasions and cloudy on the other two. And so there was hope, it being a summer month, that the weather would be clear on eclipse day.

On a different note, months earlier the United States Naval Observatory had prepared a thirty-page pamphlet that explained to both professional astronomers and novice observers how to record the eclipse. For those who intended to sketch the eclipse, it recommended that they use a standard format consisting of a sheet of paper nine inches wide and twelve inches long with a black disk drawn on it beforehand, the disk should be one and a quarter inch in diameter to represent the eclipsed Sun. It also suggested—as Langley had wished he had done—that a person who planned to do a sketch be blindfolded beforehand and that someone should be standing nearby to notify when totality began. Then, when the blindfold was removed, the sketcher should first note the color of the corona, whether it formed a concentric ring or was diffuse and uneven around the Sun. The positions and shapes of red prominences should be noted. The entire corona should then be sketched quickly, and luminous patches or dark rifts should be included. Then details of the corona should be added, whether there were distinct rays and which rays were radial to the Sun and which had curved trajectories. Above all—and this would be difficult for almost any artist to follow—the instructions stated that the work should not be done so hurriedly as to deprive it of all value. And never, the instructions continued, should an original sketch be altered after totality is over. Instead, additional drawings should be made from memory.

Hundreds of people did follow the instructions. Their work resides today in several boxes in the Old Military Records Division of the National Archives building in College Park, Maryland. Most of the drawings are done in pencil, some in ink, a few in chalk, and one is in oil. A dozen or so of the drawings were completed by college students who were visiting Denver and who seated themselves on Capitol Hill on the day of the eclipse. To aid in their work, they were supplied with opera glasses and binoculars, courtesy of the Chicago Astronomical Society that had borrowed them from the owner of a Chicago theater.

Celebrations could be found in towns all along the eclipse path. In Colorado Springs a local hotelier hired a band to play Beethoven during the eclipse. Nearby in the picturesque Garden of the Gods, a region of spectacular red-rock formations, tents were set up and a piece of smoked glass was handed to everyone who attended for a fee of twenty-five cents. Wooden platforms were erected in public squares so that people could gather to view the eclipse. Banks were shut down and stores closed. Newspapers published special "eclipse editions." John Emerson, a miner who was excavating a promising lode he had discovered a few weeks earlier near Leadville, Colorado, came out of the ground long enough to look up in the sky and marvel at the eclipse.

A party of men and women of the Central Presbyterian Church of Denver left on 7:00 A.M. on the morning of the eclipse. They traveled by train, then by carriage, and finally by horseback to Argentine Pass along the Continental Divide and forty miles west of Denver. From that position, just before 4:30 in the afternoon, those with a keen eye could see the Moon's shadow approach from the northwest. One by one they saw lofty peaks—Longs Peak and the Mount of the Holy Cross—covered in darkness. Then they too were enveloped, though far to the south, Pikes Peak was still in sunlight. But in a few seconds it too succumbed. After two and a half minutes, they saw the reverse as sunlight reappeared over Longs Peak, then across Argentine Pass and, finally, at Pikes Peak.

Those at Argentine Pass probably wondered what it was like to view the eclipse from Pikes Peak, the highest summit in that range of the Rocky Mountains and already one of Colorado's most famous landmarks. Langley would reveal all in his official report.

Up on the peak, he had originally planned to keep himself in darkness until the moment of totality, but the necessity of helping others in the group prepare for the eclipse prevented him from doing so. And so his eyes were not adjusted to darkness when the eclipse came. And he regretted it.

His first impression was that the corona was not as bright as the one he had seen in 1869. Right around the circumference of the Sun it appeared as a narrow ring—hardly more than a line—of vivid light. Beyond that it faded rapidly into a nebulous halo composed of subtle streaks extending away from the Sun. The distance the streaks extended was highly uneven. Most kept close to the Sun, but on opposite sides there were two great streaks—Langley describes them as "rays"—that resembled long wings that Langley could follow far from the Sun. It was for this that he regretted his eyes were not adapted to the darkness, because if they were he was sure he would have seen each wing extend across most of the sky.

Maria Mitchell also viewed the eclipse. She was with six of her students—"all good women and true," she would write. They took up a station just north of the Sisters' Hospital near Denver. "Our camping place was near the house occupied by sisters of charity, and the black-robed, sweet-faced women came out to offer us the refreshing cup of tea and the new-made bread." They set up three telescopes, including the Dollond. She assigned one student to assist her and take notes. Two other pairs of women were assigned to the other two telescopes. The remaining woman sat nearby and counted out the seconds so that the other three pairs of women could hear.

For the few minutes immediately before the eclipse, Mitchell imposed absolute quiet on the group. They too saw the shadow approach from the northwest. "It was the flitting of the cloud shadow over hill and dale," Mitchell would recall. Instead, it was as if the sky had dropped a sheet of darkness over them. Then it was two minutes and forty seconds of intense work, recording the positions of red prominences—"so many, so brilliant, so fantastic, so weirdly changing"—and the shape of the corona, Mitchell also seeing the two long wings, which she pronounced "the great glory."

Before 1878, there is a noticeable lack of eclipse descriptions written by women, an exception being the one Mitchell provided of the 1869 eclipse. But for 1878 there are several. The description given by Angeline Hall, the wife of Asaph Hall, was published as part of the official government report about the eclipse.* She, too, mentioned the long rays and the "soft white" color of the corona. Mary Eastman was the wife of John Eastman, also an astronomer at the Naval Observatory. She was allowed entry to the Masonic Hall in Las Animas, Colorado. She locked herself in so that she would be absolutely alone and undisturbed. She found a room on the second floor that had a window that looked in the direction of the Sun. As totality neared, she noticed that shadows of buildings deepened to an inky blackness. Minutes before totality, she had "a general visual sensation" as the remaining feeble sunlight seemed to flicker. Then, with seconds to go, the corona flashed into view and what, in her words, seemed to be "a very rapid withdrawal of sunlight." She carefully scrutinized the corona with field glasses, seeing the corona as two great rays stretched out away from the Sun. Close to the Sun were wisps of light, some slightly twisted. Turning her attention back to the horizon, the land beyond the edge of the shadow, a distance of more than fifty miles was tinged orange, as in a sunset in calm weather. The shadow itself seemed like a great lid moving across the landscape. Turning back to the Sun, sunlight flashed out almost instantaneously. In her report, also published by the Naval Observatory, she was careful to note that it had been written "without consultation with anyone, and from notes taken immediately after totality, before I had seen, heard, or spoken to any person whatever."

Mary Draper might also have provided a lively description of the eclipse, but she was a member—in fact, the only female member—of

* Angeline and Asaph Hall were an interesting couple. He learned his college mathematics from her when she was an instructor at New-York Central College in McGrawville. She was an ardent suffragist, often exchanging letters with Susan B. Anthony. He was an equally passionate anti-suffragist. She occasionally worked with him at the Naval Observatory doing astronomical calculations, that is, until they had an argument whether women should receive the same pay as men for doing the same type of calculation. She was for it; he was against it. She never helped him with mathematics again.

an expedition led by her husband Henry, son of John Draper who, many years earlier, had taken the first photograph of the Moon. Mary and Henry were an astronomical duo, observing and photographing the night sky from their home in Hastings-on-Hudson, about twenty miles north of New York City. In fact, on the day of their wedding, after the ceremony, they went to New York City to choose a mirror for a new telescope they planned to build. During their nighttime work, Mary prepared and developed the wet-collodion plates while Henry operated the telescope and took the photographs. It was said that neither ever observed without the other.

In 1878 they went to Rawlins, Wyoming, a frontier railroad town of about eight hundred persons, to photograph the eclipse, arriving with two other men who were also members of the expedition, chemist George Barker of the University of Pennsylvania and Henry Morton, President of the Stevens Institute in Hoboken, New Jersey. Each person had an assignment on the day of the eclipse. Morton would align the telescope. Henry Draper would operate the camera. Barker would prepare the wet-collodion plates. And Mary Draper would count out loud as each second passed during the eclipse.

This was an era when many people thought there were inviolable differences in men and women. Men were assertive and rational; women, by their nature, were passive and emotional. Because an eclipse expedition succeeded only if everyone did an assigned task and no one was distracted by the unusual views, it was thought best that Mary Draper spend her time sitting inside a tent "lest the vision might unnerve her." And that is what happened. After years of working closely with her husband, and after traveling hundreds of miles and knowing that the weather was clear, all she ever experienced of the 1878 total solar eclipse were the few minutes of darkness that surrounded her.

For many astronomers—Hall, Newcomb, Langley, Mitchell—the event in 1878 would be the last total solar eclipse they would see. The next such event would not cross the United States again until 1889. By then, these pioneers of American astronomy were in their later years and unwilling to cope with the challenges of travel.

But, as always, there was a new generation who were eager to see the world. And from them came a new type of traveler—the eclipse chaser. And the most famous eclipse chaser of the era was not an astronomer. She was Mabel Loomis Todd, one of the era's most popular public lecturers and the wife of the astronomer at Amherst College in Massachusetts.

———•———

The crucial date in her life was August 31, 1881. On that day, Mabel Loomis Todd and her husband of two years arrived in Amherst in rural Massachusetts. She was twenty-four, full of spirit and ambition, and fresh from a vibrant social life in Washington, D.C., where she had been raised and where, as she recorded it, she had been brought up "among distinguished scientists and men of letters." By contrast, Amherst was a quiet college town. Its population was a few thousand. The faculty was mostly elderly men; their wives were ladies of quiet tastes who dressed in dark colors. Dinners were at six. Few people danced or sang. As Mabel Todd wrote in her journal on the day she arrived in Amherst: "What have I done?"

She was there because her husband, David Todd, had accepted a position as professor of astronomy at Amherst College, lured there by college officials who promised him a new telescope—a promise that did come true twenty-five years later. In the meantime, he used the small telescope that was then on hand. She immediately made her way into Amherst society. And the center of Amherst society was the Dickinson household.

Austin Dickinson was a gargantuan man always impeccably dressed and with a shock of copper-red hair. He was the leading lawyer of the region and a trustee and treasurer of Amherst College. On September 10, 1882, more than a year after they arrived, the Todds were invited to the Dickinson home. They were introduced to Austin's wife, Susan, and to his sister Lavinia. During the evening, Mabel was asked to play the piano and sing. At the completion of her impromptu recital, a maid presented her with a glass of sherry and a poem written by Austin's other sister, Emily Dickinson, who was sitting in another room.*

"She writes the strangest poems & very remarkable ones," Mabel Todd would write later. It was Mabel who first shaped how we perceive Emily Dickinson today. Todd wrote to her mother:

> I must tell you about the *character* of Amherst. It is a lady whom the people call the *Myth*. She is a sister of Mr. Dickinson, & seems to be the climax of all the family oddity. She has not been outside of her own house in fifteen years, except once to see a new church, when she crept out at night, & viewed it by moonlight. . . . She writes finely, but no one *ever* sees her.

For years Todd would visit the Dickinson house and play the piano and sing, and notes would be passed between her and Emily. Occasionally, they would have a conversation, and Todd became familiar with her voice, but she was always confined to a drawing room while the recluse poet stood in a darkened hallway just beyond where she might have been seen.

There is so much more to the story between the Todds and the Dickinsons. Mabel Todd and Austin Dickinson became passionate lovers, a secret that everyone in Amherst seemed to share. The notes they passed between themselves and the letters they sent to each other when one was away are still able to steam up the most innocent mind. David Todd, too, had his infidelities, some possibly involving pubescent girls. The Todds were tolerant of each other's behavior. What Austin's sisters, Lavinia and Emily, thought of these relationships is not clear because both remained reclusive, Emily most of all. Susan Dickinson did object, though there was not much that she could do, except for one night when she scraped the wallpaper in one room of the house so deeply with her fingernails

---

* Though it is not certain, several scholars have suggested that the poem sent to Mabel Todd that night began: "Elysium is as far as to/The very nearest Room."

during a tantrum in front of Austin that the wallpaper had to be replaced.

In May 1886 Emily became ill, and Mabel spent much of her time at the Dickinson house concerned about her neighbor. On May 15 Mabel wrote in her journal: "Emily just leaving. A few very sad minutes." And then the poet was dead.

The funeral was held four days later inside the Dickinson house. Only a few people were invited. Mabel was one of them, escorted by Austin. As one of the neighbors who attended recalled, Mrs. Todd was "dressed in black—looking haggard as if she had lost a dear friend." It was the first time she had ever seen Emily's face, which she recorded simply as "beautiful" and "poetical." Some who watched the funeral procession from afar would remember how small the coffin looked.

In many ways, Emily Dickinson's death marked a turning point in Mabel Todd's life. She and Austin continued to be devoted lovers, although as Austin continued to grieve and became less interested in life, Mabel turned to her husband. It was not for sexual satisfaction but for a different reason, one that may be used by someone who is coping with a sudden loss. Mabel and David Todd started to travel. And the first trip they took was to Japan to see a total solar eclipse.

They left Boston by train on June 9, 1887, traveling to Japan the fastest way possible, first along the Canadian Pacific Railway to Vancouver, British Columbia, then by ship to Yokohama, arriving on July 8, six weeks before the solar eclipse. After much consideration—how easy was it to transport the equipment and where was the weather most likely to be favorable—David Todd chose the village of Shirakawa on a broad plain about a hundred miles north of Tokyo for his observing station. He erected the telescopes and cameras and tested them repeatedly. Mabel Todd's role would be to sketch the corona, to try to capture the great wings that had been seen during the 1878 eclipse. But, as her husband later wrote in his official report, "the purely eclipse-results of the work at Shirakawa were disheartening in the extreme."

The total phase of the eclipse would occur at about three o'clock in the afternoon. By two o'clock the sky was still perfectly clear. A half hour later a slender finger of cloud could be seen rising in the west. The cloud started to spread eastward over Shirakawa. The source was the eruption of a volcano twenty-five miles away. Mabel Todd also saw the threatening cloud as she assessed the situation, "quietly and simultaneously our 'massive enemies' collected." After another half hour the entire sky was covered. Totality came and the sky darkened. "Silence like death filled . . . the town and all the country round," she wrote. "Not a word was spoken. Even the air was motionless, as if all nature sympathized with our pain and suspense." Her thoughts went back to the 8,000 miles they had traveled over land and stormy sea and to "the ton of telescopes brought with such care, the weeks of patient work." They had trusted nature, Mabel wrote, and nature had failed them.

After they returned to Amherst, Mabel received a surprise. Soon after Emily Dickinson's death her sister Lavinia had found a box containing forty of Emily's poems written on folded sheets of stationery and bound in a score of small hand-sewn books.* Believing, though not knowing, that her sister had talent, Lavinia thought the poems should be published. She first appealed to her sister-in-law, Susan, to edit and arrange the poems, but Susan seemed unable to get started. Next she asked published poet and essayist Thomas Higginson, who had exchanged letters with Emily for many years. He replied that he was too busy to take on the difficulties of Emily's poems and, furthermore, that he doubted enough suitable ones could be found to fill a volume. As her last hope, Lavinia approached Mabel. Would she edit Emily's poems?

At first, she too was reluctant. She was an author herself, having published the novel *Footprints* in 1883, a story about a young woman who falls in love with a quiet, lonely man. (A comparison with her

---

* During Emily's lifetime, only seven of her poems were published, and those anonymously.

relationship with Austin who was twenty-seven years older than Mabel is unmistakable.) And so she knew of the difficulty of preparing a manuscript for publication. At the time Lavinia approached her, she was writing magazine articles about her recent trip to Japan and about eclipses. She also thought the unconventionality of Emily's poems might repel publishers. But she was strangely drawn to them. In the end, she agreed. She also convinced Higginson to work with her on the poems.

For the next several years she devoted three or four hours each morning to transcribing poems using a typewriter. Lavinia continued to discover more; some mornings she would bring a basket filled with new poems to the Todd household. Somehow Mabel also found time to write additional magazine articles and a monthly book column for *Home Magazine*. She also traveled throughout New England giving lectures about astronomy as well as Emily Dickinson and her poetry. By 1896 she completed three volumes of poems and a collection of letters written to and by the poet. She also completed a popular science book, *Total Eclipses of the Sun*. It was at the end of this frantic period of editing and writing and publishing that Mabel Todd resumed chasing solar eclipses.

On August 9, 1896, she and David were in Hokkaido, Japan, and failed to see the corona because of clouds. Their next trip was successful.

They saw the solar eclipse in the city of Tripoli in North Africa on May 28, 1900. Again a ton of telescopes and cameras was brought to photograph the eclipse. The equipment was set up on the roof of the official residence of the British consul-general in Tripoli. For days beforehand, assistants were found and trained in how to count the time and how to watch and record Baily's beads. On the morning of the eclipse Mabel awoke from a frightful dream about fog and storms to find a morning sky of unsurpassed clarity.

The day quickly grew hot. There was a concern that the telescope and camera mechanisms might not work in such heat. The glare of white buildings was almost blinding.

With minutes to go before totality, Mabel looked across the city. People were on every rooftop. Even the minarets were occupied. Crowds had gathered in the streets.

Her job was to watch for shadow bands. A full ten minutes before totality, with sunlight coming from only a thin crescent, she could see thin waves of light and dark run across the landscape, moving faster than a person could walk. Just as the corona appeared, an instant hush rolled over the city. After another instant people on rooftops, in minarets, and in the streets below all raised their hands toward the Sun, and Mabel could hear them pray. She grabbed a sketchbook and drew the corona weaving together lines to represent individual filaments. The darkness lasted fifty-one seconds. Then "a needle-shaft of true, returning sunlight flashed over the world." Totality was over.

She summarized her feelings afterward:

> I doubt if the effect of witnessing a total eclipse ever quite passes away. The impression is singularly vivid and quieting for days, and can never be wholly lost. A startling nearness to the gigantic forces of nature and their inconceivable operation seems to have been established. Personalities and towns and cities, and hates and jealousies, and even mundane hopes, grow very small and very far away.

———•———

The Todds were not the only eclipse chasers of the era. Norman Lockyer, the man who discovered helium, managed to organize and travel on eight eclipse expeditions. William Pickering of Harvard College Observatory went on five. William Campbell, the director of Lick Observatory in California, organized nine expeditions including one to Russia in 1914. In each case, about half of the expeditions the weather was clear and the corona was visible. Charles Abbot of the Smithsonian Institution had an amazing record of seeing six total solar eclipses and missing only two because of bad weather. But the Todds were supreme. Mabel Todd, in addition to writing and giving public lectures about eclipses—she made enough money to underwrite an eclipse expedition back to Tripoli in 1905—went on seven expeditions and saw the corona three times. David Todd went on twelve expeditions and had good weather during seven of them. His record would not be exceeded by anyone until the 1970s when jet travel became widely available.*

And there was much else accomplished from these many expeditions beyond photographs of the corona and an eclipsed Sun. Mabel Todd was a collector of art and artifacts and much of what she collected during her travels was donated to the Peabody Museum at Yale University and to the Peabody Essex Museum in Salem, Massachusetts. Included were baskets and pottery from Japan, and garments and ornaments from Tripoli. She also wrote extensively about her travels to Asia, Africa, and South America. And she accomplished a number of firsts along the way. She was the first Western woman to reach the summit of Mount Fuji in Japan and was also one of the first people to be permitted to photograph women in a harem.

A close reading of expedition accounts written by her, her husband, and many others reveals something else—a deep interest in the folklore of eclipses and how people of different cultures responded when the Sun suddenly went dark in sky.

---

* The current record for the number of total solar eclipses witnessed by an individual is held by Jay Pasachoff of Williams College, Williamstown, Massachusetts, who has seen 33 and who has stood in the Moon's shadow for an accumulative time of 1 hour 22 minutes 17 seconds.

CHAPTER TEN

CHAPTER TEN

Keys and
Kettledrums

*The Sun and the Moon are brother and sister, guilty
of incest.*

—from a Cherokee myth

An eclipse is never just about science. Nor it is simply about the spectacular and unusual views or the strange phenomena. The excitement of an eclipse comes mainly from how one reacts to an eclipse, that is, to the sudden disappearance of our most reliable and stable source of light and heat—the Sun.

I confess that I am always a bit anxious during the minutes just before an eclipse begins. Though I am well aware that I live in an age when eclipses are precisely predictable, I still waver just before the Moon starts to creep in front of the Sun, wondering if the event will actually happen. And so I understand why people long ago were shocked and confused when the Sun started to disappear, why they thought that some great force must be at work, and why they knew they had to take some kind of action.

The most common explanation offered by people in prehistory for the disappearance of the Sun, or the Moon, was that it was being consumed by some great cosmic demon. To the Cherokee of North America, it was a great frog that was swallowing the Sun. In Egypt it was a serpent. To people who lived in the jungles of South America it was a jaguar they feared. In Australia a giant black bird flying in front of the Sun was the cause of a solar eclipse. To people who lived on islands off the coast of Venezuela, it was a devil in human form known as Maboia.

These are only a few of the many examples that could be cited. Whether such explanations were accepted as literal truths or as metaphors is a question for cultural anthropologists and students of religion to answer. Here the important point is that the manner in which people respond to an eclipse was almost universal and consisted of making loud noises to drive away the demon.

And this is how it has been throughout the world and throughout most of time. The Roman historian Tacitus tells us that when those who were conspiring against Emperor Tiberius felt frustrated in their efforts after seeing a lunar eclipse, they resorted to clanging brass and blasting trumpets to scare away the darkness. In ancient China women tapped loudly on bronze mirrors during a lunar eclipse to chase away a mythical dog that was said to be devouring the feminine Moon. Bernardino de Sahagún, a Spanish priest writing in the sixteenth century just after the European conquest, described the pandemonium that broke out among the Aztecs during a solar eclipse, "the common folk raising a cry, lifting their voices, making a great din, calling out, shrieking." James Adair, who resided for forty years among the Chickasaw in what is now the southeastern United States, recorded in his famous *History of the American Indians* that during a lunar eclipse in 1736 that people fired guns, shouted, beat kettles, rang bells, and made "the most horrid noises that human beings possible could." More recently, in 1917, the British army officer T. E. Lawrence, known better as "Lawrence of Arabia," took advantage of a lunar eclipse to lead Bedouin troops on a successful siege of the city of Aqaba, knowing the defenders would not notice their approach because they would be preoccupied with banging pots, shooting guns, and making other loud noises.

I have always thought that the way people react to eclipses was a visceral response that originated from somewhere deep inside the most primitive part of the brain. I say this because I am often stunned to silence when seeing an eclipse. At other times, I seem unable to stop repeating to myself that what I am witnessing is a natural phenomenon that is easy to explain. My reaction seems unreasonable and irrational. And, yet, I am not alone. Others have expressed similar feelings.

To quote two contemporary writers who have sought out eclipses, Annie Dillard, who witnessed a total solar eclipse in 1979, described the event as being "like dying" and felt that "the world was wrong." To her, totality came when "the sky snapped over the sun like a lens cover." At the same moment, "the hatch in the brain slammed." Dava Sobel was on a cruise ship off the coast of Mexico in 1991 when she enjoyed seven minutes—almost the longest possible— within the Moon's shadow, remembering that the visions and the sensations brought "such a lump in my throat that I have to will myself to breathe normally." She was holding the hand of a companion at the time and could feel it shake. A much more succinct account was provided by veteran sun-watcher Jack Zirker, director of a solar observatory at Sunspot, New Mexico, who was in India in 1980 for a total solar eclipse. He set up a motorized camera to take a series of exposures and was going through a well-rehearsed procedure when he was startled by what he saw in the sky, describing the scene in his words, "Incredible! It is the eye of God."

A less emotional and broader-based, more clinical analysis of why people behave irrationally during eclipses was offered by French anthropologist Claude Lévi-Strauss. In his groundbreaking book, *The Raw and The Cooked*, published in 1964, in which he argued that there was no fundamental difference between the mind of a civilized person and that of a so-called primitive, he showed how the reaction to an eclipse related to other human activities. Specifically, he associated the making of loud noises during an eclipse to how people in many societies react after the consummation of an unseemly marriage.

He gave as examples marriages between people who have a great disparity in age, who marry for social advancement or for money, who marry foreigners or who do not respect the prohibitions of kinship. Such unions are seen as dangerous to the welfare and stability of society. According to Lévi-Strauss, the making of derisive noise outside the house of such a newlywed couple was an unofficial way to show displeasure at the union—a common practice in that time. And he demonstrated this by recounting a story common to many societies around the world that related incestuous behavior to eclipses.

The story was told throughout most of the Americas, from southern Brazil to the Bering Strait, and across northern and eastern Asia and the Malay Archipelago. It is a story about the origin of the Sun and the Moon. Though the details vary depending on where the story was told, it was basically about how a brother pursued a sister, who finally took refuge in the sky and became the Sun. Undeterred, he became the Moon. Each month as they neared each other, he tried to rape her. The blood color of a total lunar eclipse was evidence of a recent rape. In retribution, she rubbed his face with a horrible juice that stained his face, accounting for the dark spots on the Moon. And, just as it is the duty of people to show their displeasure with an unseemly marriage on Earth by noisemaking, so it is also their duty to stop the incestuous rape that is occurring in the sky during an eclipse. At least, that is how Lévi-Strauss made the connection: Incestuous intercourse *is* an eclipse and an eclipse *is* a cosmic attempt to break social order.

Whether one agrees with Lévi-Strauss or not, people today do follow rituals that they believe will protect them from the evil influences of eclipses. Among the more benign is the turning over of pots and buckets and the covering of water wells—a custom practiced in such widely different places as Japan, the Yukon, and southern Germany—to prevent poison from raining down from the Sun or the Moon and contaminating the contents. Another is the gathering of herbs during the darkness of a lunar eclipse, the somber red light somehow adding potency to plants. Shakespeare mentioned such ingredients used by the three witches when making a brew in *Macbeth*, which included "slips of yew slivered in the moon's eclipse."

One of the most common rituals is the protection of pregnant women. Throughout much of Latin America, including parts of the southern United States, a metal key or a safety pin is attached to clothing near the abdomen as protection from an eclipse-caused abortion. The metal key or pin also protects the unborn child from developing a cleft lip, a deformity of a baby's mouth taken by the "bite" seen when the Moon passed in front of the Sun. This ritual has its roots among the Aztecs (and possibly much earlier), who used pieces of obsidian placed in a pregnant woman's mouth or into her girdle to protect against eclipses. In modern times, to ensure the safety of an unborn child, in a house of a pregnant woman, someone will fill a pot with water, then place metal scissors, opened in the shape of a cross, at the bottom of the pot.

There is an interesting contrast between people of the South Pacific and those of ancient Rome whether one should have sexual intercourse during an eclipse. To Pacific islanders the exact meeting of the Sun and the Moon in the sky was the equivalent of cosmic copulation, a reenactment of creation itself when stars suddenly appeared. And so everyone should join in the reenactment. Conversely, according to Roman naturalist Pliny the Elder, who assembled the world's first encyclopedia and who set the tone for Western ideas for more than a millennium, having sexual intercourse during an eclipse was the worst thing a person could do. As he wrote: "Congress with a woman at such a time being noxious, it would end with fatal effects to a man." It should be mentioned that Pliny also wrote that if a woman who was menstruating walked around a field of wheat, caterpillars, worms, beetles and other vermin would fall from the stalks. And that touching a woman who was menstruating provided a cure for gout or for the bite from a rabid dog.

In India self-immolation was practiced in ancient times during eclipses as a means to achieve salvation. As evidence, in South India there are more than a thousand stone inscriptions dating back to the third century B.C.E. that record the practice. In modern India, during a total solar eclipse in 1995, more than a half million devout Hindus entered sacred lakes and rivers, bowing their heads and praying toward the eclipsed Sun, heeding an adage in the Hindu epic *Mahabharata*: "Salvation is verily his who bathes or gives alms at the time of a solar eclipse."

One should not have the impression that eclipse rituals were minor, short-lived affairs. They could last for days and be quite elaborate.

Those performed for lunar eclipses in ancient times at the city of Uruk in modern-day Iraq consisted of two parts, one ritual done during the eclipse and another the next day, involving not only priests and royalty but the populace as a whole. They were rigidly prescribed to include specific prayers, long lamentations, and sacrifices. Large bronze kettledrums were removed from a storehouse and set up near the city gate to be played by one group of priests. Another group started a massive fire, also near the city gate, which would be kept burning for the duration of the eclipse. Meanwhile, a third group of priests would repeat specific phrases asking the gods not to burden the city with pestilence or famine or bring bad weather. Ordinary people were encouraged to assemble, though each person must have a headdress that could be removed only after specific prayers were said. Soldiers were also involved, and in one account their bodies were covered with a special mud and their swords were hung from slings on their backs. As soon as the Moon was free of the Earth's shadow, the playing of the kettledrums was stopped and the fire extinguished, the cold embers carried to a nearby river where more incantations were offered to the gods. And the day after the eclipse this was repeated to ensure that the city and its inhabitants would be safe.

And such elaborate performances and rituals were not restricted to only highly structured urban civilizations. People who lived in small tribes along the Bella Coola River of British Columbia in Canada performed a three-day ceremony that began with the blackening of faces to represent the eclipsed Moon, and that was presided over by a woman who represented the Moon's spirit. It was one of the most sacred ceremonies performed by these people. And if one of the tribal dancers refused to participate, death was a likely punishment.

As a final example, the Serrano, an indigenous people who lived in the valleys of the San Bernardino and San Gabriel mountains of southern California, had an eclipse ritual that lasted days. It began when the first person who saw an eclipse of either the Sun or the Moon raised a shout that was taken up by everyone. People then gathered in a ceremonial house where tribal shamans sang and danced. Anyone could join them. Because the Serrano believed eclipses were caused by spirits of the dead, or by the spirits of those who would soon die, while they danced, the shamans also watched individuals in the crowd, looking for those whose spirits were leaving them. The shamans then decided which spirits should be allowed to leave, informing those unfortunate people that they would soon die. As in other cultures, the shadow of an eclipse was thought to contain poison, and so food could not be touched during the eclipse. After it was over, people cleansed themselves by bathing, then drank concoctions made from boiling a pre-scribed herb to cleanse themselves further, and finally everyone ate together, signaling a breaking of the fast.

––––•––––

Humans are not the only creatures to react to eclipses, although we may certainly be the noisiest. The fact that birds respond to total solar eclipses has been known for a long time. One of the earliest accounts comes from the Czech astrologer Cyprianus Leovitius, who in 1544 saw that "darkness was coming on as if in evening twilight and the birds, which had been singing, became mute." It was an observation that has been repeated many times, revealing birds' sensitivity to changes in the light. It was a lesson quickly learned by American inventor Thomas Edison.

Edison was part of the eclipse expedition sent to Rawlins, Wyoming, in 1878. The previous year he had risen to national fame for his invention of the phonograph. He now wanted to do something equally notable for a solar eclipse. And so, with his considerable inventive powers, he devised a new type of scientific instrument, the "tasimeter," from the Greek for "to measure anything." Its purpose was to determine the temperature of the Sun's corona. Edison finished it just two days before leaving for the West.

It was a delicate instrument, and Edison sought a quiet sheltered place to set it up. On the day of the eclipse, he found a vacant shed with a doorway that would give an unobstructed view of the eclipse. He waited alone inside the shed watching as the Moon slowly crept over the Sun. Just as darkness began and he was ready to take the first measurement, to his surprise, a flock of chickens rushed into the shed, flying in, around, and over the frantic inventor and disrupting his work. The shed he had chosen was a chicken coop and, seeing the impending darkness, its occupants had come home to roost.

A wide variety of animals have been watched before and during total solar eclipses. Crickets begin to chirp during totality, then stop when sunlight returns. Bees cease the gathering of honey and head back to the hive. Those that do not reach the hive when darkness descends keep flying around or land in the grass until it is light again. Ants busily carrying their burdens stop and remain motionless until totality ends, then resume their work. Katydids, dragonflies, butterflies, and gnats land, their flights resuming when darkness ends.

Frogs begin to croak. Toads spring into action and hop around searching for insects and worms. Then, when the Sun appears again, they resume their ordinary daytime quiet. Even microscopic plankton responds to total solar eclipses, rising close to the water's surface, just as they do during evening twilight.

Brook trout, goldfish, small-mouth bass, and white perch are just a few of the types of fish that have been observed. Trout, bass, and perch respond to a solar eclipse in the same way they do as night-fall approaches; they stop feeding and go to the bottom. Goldfish, however, seem to have no response to darkness.

But by far, the behavior of birds is the most widely reported. As Edison discovered inadvertently, chickens go to roost. Turkeys, ducks, and geese settle onto the ground wherever they are. Nightingales and warblers stop singing. Owls and nighthawks begin to hunt. Those in captivity such as canaries and parrots are uneasy, suggesting they are nervous or bewildered. An intensive study of bird behavior involving scores of observers was done during the 1932 total solar eclipse visible from Massachusetts, New Hampshire, and Maine. Those observers showed that birds do not react to solar eclipses unless 98% or more of the Sun is covered.

In general, dogs seem to pay no attention to the darkening conditions, nor do cats. Also no reaction has been seen in seals, rabbits, deer, or sheep, and none in rats, gerbils, or field mice. Bats that are settled in trees leave their roost and hunt for insects, while those in caves, to no surprise, are unaffected. Grazing cattle do start walking back to the barn as the sky noticeably darkens. And horses have given no sign whatsoever that an eclipse is happening. Primates, however, do react.

Howler monkeys start to howl, apparently thinking that sunset is near. Rhesus monkeys stop whatever they are doing and huddle together. Chimpanzees have what can only be described as a thoughtful response.

A captive group of sixteen chimpanzees—seven infants and juveniles, eight female adults, and one male adult—were observed at a primate center near Atlanta, Georgia, during an annular solar eclipse in 1994. At one point 99.7% of the Sun was blocked by the Moon, which left a ring of bright light around the periphery. About ten minutes before the maximum eclipse the infants and juveniles and most of the adult females moved to the top of a climbing structure. As the eclipse progressed, the others joined them. As the group sat there, they oriented themselves to face the eclipsed Sun. At one point one of the juveniles stood upright and gestured in the direction of the Sun and the Moon. As the eclipse ended and sunlight became brighter, they began to descend. Ten minutes later all individuals were down from the climbing structure once more.

Such behavior by the chimpanzees had not been observed prior to or after the eclipse, nor had it ever been observed at the Atlanta primate center during familiar darkness that approaches at sunset. And that gives us pause. How exactly did our ancient ancestors react? We will never know, but one can receive a hint when one stands within the Moon's shadow and looks up and senses a strange feeling of uneasiness.

—◦—

How plants react to the sudden darkness of solar eclipses is still an open question. The flowers of anemone, gentian, mimosa, and crocus have all been seen to close fully. In one case, in Sweden, during the total solar eclipse of 1851, a night violet was noticed to release its nocturnal scent.

—◦—

A few solar eclipses have influenced the history of the United States, though, as might be expected, by indirect means. The first occurred in 1806. Nathaniel Bowditch, the author of the *Practical Navigator,* a compendium of essential nautical and astronomical knowledge for surveyors and ship captains, observed the eclipse from the comfort of his private garden in Salem, Massachusetts. Samuel Williams, who had led the failed attempt to view the Sun's corona from Maine in 1780 during the Revolutionary War, watched the 1806 solar eclipse from his home in Rutland, Vermont. William Cranch Bond, who was seventeen years old at the time, saw the eclipse from the rooftop of his family's house in Boston. The view of this magnificent event piqued his interest in astronomy. He would go on to become the first director of Harvard College Observatory and educated the next generation of American astronomers. This eclipse and his other stargazing so inspired him that he built a house with shutters in the ceiling of the parlor that could be opened, allowing him and his guests to study the night sky through a telescope. Notwithstanding the scientific observations made by Bowditch, Williams, and others, and the influence the 1806 solar eclipse had on young Bond, the historical importance of this eclipse happened to the West in what was then known as the Indiana Territory and culminated five years later at a place known as Tippecanoe.

The governor of the Indiana Territory in 1806 was William Henry Harrison, a former army general, who had been directed by President Thomas Jefferson to obtain control of as much land within the territory as possible. He did so through treaties that paid little or nothing for the land to the Native Americans who were already living there. As expected, that brought Harrison in direct conflict with local leaders, in particular, the Shawnee chief Tecumseh and his brother Tenskwatawa.

In 1805 Tenskwatawa, who had been an alcoholic much of his life, experienced a number of visions that caused him to urge members of his tribe and other tribes to abstain from using alcohol and to cleanse themselves of all influences of the white race. He and Tecumseh organized a community near the present site of Greenville, Indiana, where Tenskwatawa continued to preach. Sensing the growing influence of the man who was now known as the Great Shawnee Prophet, in the spring of 1806 Harrison issued a challenge: "Demand of him some proof," he wrote to those who were joining Tenskwatawa, "of his being the messenger of the Deity."

If God has really employed him, he has doubtlessly authorized him to perform miracles, that he may be known and received as a prophet. If he is really a prophet,

ask him to cause the sun to stand still—the moon to alter
its course—the rivers to cease to flow—or the dead to rise
from their graves.

Before long, Tenskwatawa seemed to have performed such a
miracle. It was said that in early June he had stood in front of his
followers and declared that on a certain day, June 16, the Sun would
darken at noon. And a total solar eclipse did take place. As the Sun
faded into an eerie twilight, the Great Shawnee Prophet shouted,
"Did I not speak the truth? See, the sun is dark!" Doubters were
now convinced of his great powers. More people joined Tecumseh
and Tenskwatawa at Greenville. And Harrison's influence lessened.
But could Tenskwatawa have actually made such a prediction?

Some historians have suggested that Tenskwatawa may have
learned about the eclipse beforehand from astronomers who were
passing through the area looking for potential sites where they
might view it. Others have noted that some of the Americans—as
well as some of the British—who were living in the area, including
Harrison and his soldiers, had almanacs and that they certainly
talked about the coming eclipse and that Tenskwatawa may have
overheard such conversations. Neither suggestion seems reasonable. First, there is no record of astronomers passing through or
witnessing the eclipse from the Indiana Territory. All of the reported
observations were made from close to the eastern coast. The farthest west any scientific observation was made was in Lancaster,
Pennsylvania. Second, if Tenskwatawa heard conversations about
the coming eclipse, other tribal leaders would have heard similar
conversations and might have been tempted to make their own
predictions. There is no evidence that other predictions were made.
But there is a third possibility.

In 1797, at age twenty-two, Tenskwatawa began to receive instructions in Shawnee methods of healing and incantations from the
aging shaman Penagashea. The instructions almost certainly
included some astronomical lore, something that is part of every
culture. Exactly what the shaman might have known about the

movement of objects in the sky is not known because Shawnee knowledge was never written down but was passed down orally. But it is not a stretch to assume it included eclipses and the ability to predict them by simply counting combinations of 177 and 178 days.

Penagashea died in 1804 and Tenskwatawa took over the role of shaman. Here it is important to note that a partial solar eclipse was visible from the Indiana Territory at sunset on June 26, 1805. Tenskwatawa could have used that date, then counted 355 (= 177 + 178) days, to know that a solar eclipse was possible on June 16, 1806. That the solar eclipse would be a total one would not have been known to Tenskwatawa; instead, that fact just added greatly to the drama of the prediction, making the prediction more memorable. And it made him and his brother more powerful.*

Over the next few years, more people joined Tenskwatawa and his followers. They moved several times, eventually establishing a large settlement near the juncture of the Wabash and Tippecanoe rivers near present-day Lafayette, Indiana. It was the largest community of Native Americans ever assembled in the Indiana Territory. And Harrison decided he had to act.

Tecumseh was away when Harrison led an army of almost a thousand men against Tenskwatawa and his followers. The Battle of Tippecanoe occurred on November 7, 1811. It was not the military victory that Harrison later claimed. His loss was more than forty killed and twice as many wounded. The number of Native American casualties is still a matter of debate, but it was less than the United States forces. And it did disperse the people at Tippecanoe and it ended the influence of the Shawnee Prophet. And it was used as a platform for Harrison to run for president and be elected in 1840 using the slogan "Tippecanoe and Tyler too" to remind people of the battle.

---

* It should be noted that both Tecumseh and Tenskwatawa were near Greenville, Indiana, at the time of the eclipse, which meant they were about twenty miles south of the path of totality, and so they and their followers did not see a total solar eclipse.

The second example of an eclipse influencing United States history occurred in 1831 and involved a slave rebellion. Three years earlier, while working in his owner's fields in Virginia, Nat Turner "heard a loud noise in the heavens," then saw a "Spirit" appear that told him to fight against evil. From that moment, Turner was convinced that he "was ordained for some great purpose in the hands of the Almighty." But what was the purpose and when would it happen?

The next sign came on February 12, 1831, when an annular eclipse darkened the sky. Turner saw the eclipse as "white spirits and black spirits engaged in battle." He now confided his vision to three other slaves and they conspired to act and free themselves. Turner and the others chose July 4, Independence Day, but Turner was ill, and so it was postponed. Another month passed. Then a "sign appeared again." It was August 13, a Saturday. One person who was in Philadelphia described the western sky "as one vast sea of crimson flame, lit up by some invisible agent."* After another week, early on the morning of Sunday, August 22, Turner and other slaves began a two-day slaughter of every white person they found. The rampage ended with the deaths of about sixty white people. In retaliation, white mobs and militia killed more than two hundred black people.

Turner was captured and put on trial. It was then that he dictated his famous confession to his court-appointed counsel, Thomas Gray, saying that it was the eclipse of the Sun the previous February that had caused him to "arise and prepare myself, and slay my enemies with their own weapons."

---

* An unusual sky was seen across the eastern United States in mid-August 1831. From Albany, New York, the evening sunset was tinged a deep red. In Sandusky, Ohio, the western sky looked dusky with a slight greenish cast. People in Macon, Georgia, thought another solar eclipse might be happening and grabbed pieces of smoked glass and frantically thumbed through almanacs looking for a prediction of the eclipse. The source of the disturbance in the atmosphere is still unknown, though it might have been an ash cloud erupted from Mount St. Helens, which exploded that summer.

Though Turner was soon executed, his rebellion did have a lasting, though negative, effect. Before Turner acted, some state legislators in Virginia had denounced slavery and were proposing a program of eventual emancipation. But that was over now. After Turner's rebellion, those who criticized slavery and who lived in the South either remained silent or were driven into exile, increasing further the division between the North and the South.

The third and final example is the total solar eclipse of January 1, 1889, the path of totality passing over northern California, Nevada, and territories that would soon become the states of Idaho, Wyoming, Montana, and North Dakota. By then, most Native Americans were living on reservations, usually run by corrupt and indifferent federal agents. From this system rose a prophet, the son of a Paiute shaman who spent his entire life in western Nevada, though his impact reached much further. Like many Paiutes he worked on a farm owned by a white man, a Mr. David Wilson, and so he was given the name Jack Wilson. But to history he is Wovoka, "the wood cutter."

On the day of the eclipse he was alone in his cabin near Yerington, Nevada, when he noticed the sky darkening. He went outside. The Sun was a thin crescent, almost completely obscured.* He began to have a vision of the Sun dying. In his vision, he died too. During his death, as he later described it, he saw all the people he had known who had died long ago. They were now all happy and forever young. They instructed him to return and tell the people who were still alive to stop quarreling and to live in peace. They gave him a dance to teach and told him it was to be performed until a new world replaced the current one dominated by whites. "I bring to you the promise of a day in which no white man will lay his hand on the bridle of an Indian horse," he is reported to have said. All Native Americans must dance the Ghost Dance, a means of self-purification, and most important, he said, they must "do no harm to anyone."

---

* Wovoka was about forty miles south of the path of totality. At maximum obscuration, 98 percent of the Sun was covered.

The story of his vision and of his pronouncements swept across the reservations. The Ghost Dance was performed from California to the Missouri River and from Canada to Texas. It was performed by Paiute, Shoshone, and Ute, and by tribes of the Great Plains including the Lakota. There was a concern among federal agents that this was the beginning of a military uprising, especially among the Lakota who might use the Ghost Dance as a cover for organizing a stand against further incursion onto their land by settlers. The government agents in charge of the reservations requested that federal troops be deployed to quell what they called the "Messiah craze."

It was during an attempt by troops to disarm a gathering of Lakota in January 1891 that a scuffle broke out. Guns and rifles were fired indiscriminately. At the end more than three hundred Lakota and at least two dozen soldiers were dead. The grisly encounter is known today as the Massacre at Wounded Knee.

The story of Columbus and how he used the prediction of a lunar eclipse in 1504 to exploit local people has been used as a storyline in fiction, too. The most famous example is in *A Connecticut Yankee in King Arthur's Court* in which Mark Twain has an American named Hank Morgan, who has been brought back in time, use his knowledge of when a solar eclipse is to occur to save himself from being burned at the stake. In *King Solomon's Mines* H. Rider Haggard has Englishmen in Africa predict when a lunar eclipse is to take place, thereby elevating themselves in the eyes of a local chief. Notwithstanding the Columbus story, there was such an encounter in real life before either Twain or Haggard published their stories. And it has a twist.

It is August 1869 and a physician working among the Lakota in the Dakota Territory wants to impress upon them the superiority of his knowledge in treating patients and in understanding nature. To do so, he announced the exact time, which he had taken secretly from an almanac, when the Sun would be obscured and the sky would darken. People assembled around him on the predicted day and hour. The physician, duly armed with a piece of smoked glass, which he passed around, pointed to the Sun and asked them to take a look.

It was the most favorable circumstance, the air being clear and the sky cloudless. And there was no mistake as to the time because the doctor had a pocket watch that he showed. And the Lakota seemed to give the anticipated response. At first they were passive as the sky continued to darken, with only a slit of sunlight left. Then, apparently concluding that the display had gone on long enough, they began to shout words that the physician did not understand and fire their rifles until the light of the Sun returned.

The physician, feeling he had been triumphant, addressed those around him, saying that he had demonstrated great power and that they should listen to him when he gave advice. Yes, they agreed, his power was great, but theirs was greater, because though the physician knew how to predict an eclipse accurately, they knew how to drive it away and prevent any evil consequences from happening.

CHAPTER ELEVEN

The Crucifixion and the Concorde

Thy shadow, Earth, from Pole to Central Sea
Now steals along upon the Moon's meek shine
In even monochrome and curving line
Of imperturbable serenity.

—Thomas Hardy, after the lunar
eclipse of April 11, 1903

Without a doubt the most important eclipse record that comes to us from the ancient world was made in the year 136 B.C.E. In fact, there are two written records of this eclipse. Combined they give the place where the observation was made (the city of Babylon), tell what the event was (a total solar eclipse), and give the year, month, and day when it occurred, which after conversion to our current method of reckoning time was April 15, 136 B.C.E. The two records also give the time of day when the solar eclipse happened (two hours after sunrise), and that four planets (Mercury, Venus, Mars, and Jupiter), were all visible in the sky during the darkness of the eclipse. No other record of a solar eclipse would give so many details until 1567 C.E.

And so there is no denying where the observation was made, the exact date when it was made (supported by the appearance of the four planets), and where the Sun was in the sky when it was eclipsed. And yet a serious problem was recognized with this record soon after it was discovered. If one ran the Moon backward in its orbit, taking into account all known gravitational influences—the attraction of the Earth, Sun, Venus, and Jupiter, as well as the perturbation to the Moon's motion caused by the Earth's equatorial bulge—one would discover that, yes, indeed, a total solar eclipse did occur on April 15, 136 B.C.E., but that it was *not* visible from Babylon. That is, the narrow band of totality did not pass over that city or anywhere in the Middle East. Instead, according to the calculations, that eclipse would have been seen about two thousand miles to the west, that is, it would have been seen in Morocco, Spain, and southern France, and across central Europe. Given that the eclipse *must* have been visible from the ancient city of Babylon, how might one reconcile this discrepancy? The answer came quickly.

One of two things must have happened. Either the Moon must have sped up in its orbital motion or the rate the Earth was spinning must have slowed down. In truth, both have happened—and by minuscule amounts. But because this particular eclipse happened more than two thousand years ago, the effect of a small change in the Moon's orbital speed or in the Earth's spin rate would have been multiplied many times—since 136 B.C.E. the Moon has circled around the Earth about 18,000 times and the Earth has spun completely around on its axis nearly 800,000 times—leading to a difference between the calculated and actual time of this solar eclipse of about three hours.* In addition, we know the reason why the Moon is accelerating in its orbit and why the Earth's spin rate is slowing down. Both are the result of ocean tides.

Newton in the *Principia* was the first to give a realistic explanation of ocean tides as caused by the gravitational pull of the Moon and, to a lesser extent, the Sun that gives rise to a twice-daily rise and fall of the level of the oceans. A century later, in 1754, German philosopher Immanuel Kant, whose ability to offer explanations for the natural world seemed limitless, proposed that ocean tides might have an effect on both the Moon's motion and the Earth's spin rate, though he gave only a qualitative explanation of how the interaction might work. The first detailed mathematical description was made by a French scholar, Pierre-Simon Laplace, who in 1775 not only showed how the interaction worked, but also produced the first accurate predictions of the timing and height of ocean tides. Over the next two centuries, scores of mathematicians have improved on Laplace's initial work, so that today there is a highly elaborate and highly accurate theory of ocean tides that, among other things, explains how tides affect the Moon's motion and the Earth's spin rate.

The key was the realization that, because the Earth is spinning much faster than the Moon circles the Earth (one day versus about twenty-eight days), the Earth's rotation drags the tidal bulge in the oceans so that it is slightly ahead of the spot that would be directly under the Moon. As a consequence, the gravitational attraction between the Moon and the bulge is offset from the line that connects the centers of the Earth and the Moon. That produces a torque that tends to boost the Moon in its orbit and to slow the Earth in its rotation. To say it in another way: Because the tidal bulge is slightly

---

* Three hours is the length of time for a spot on the Earth at the latitude of the city of Babylon to rotate westward two thousand miles and put the city within the path of totality of the 136 B.C.E. total solar eclipse.

ahead of the Moon, some of the Earth's rotational momentum is being transferred to the Moon's orbital momentum.

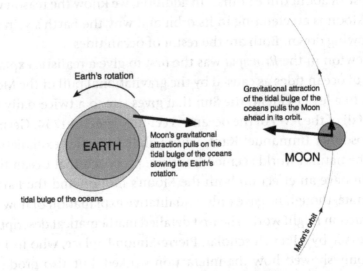

Earth's rotation

EARTH

Moon's gravitational attraction pulls on the tidal bulge of the oceans slowing the Earth's rotation.

tidal bulge of the oceans

Gravitational attraction of the tidal bulge of the oceans pulls the Moon ahead in its orbit.

MOON

Moon's orbit

The gravitational attraction between the Moon and the tidal bulge of the oceans causes the Earth's rotation to slow down and the Moon to accelerate and recede in its orbit.

By far, this is the main reason the Earth's spin rate is slowing. But there are other processes at work.

As the Earth rotates beneath the tidal bulge of the oceans, there is a slight amount of friction between ocean water and the ocean floor. The system is similar to how brakes are used to slow an automobile. A brake pad (the ocean water) presses against and creates friction against a spinning brake disk (the ocean floor). The consequence is that some of the Earth's rotational energy is lost as friction, and that causes the Earth's spin rate to slow. When looked at closely, it has been found that almost all of the rotational energy being dissipated in this manner is not occurring evenly throughout the oceans, but mostly in just three shallow seas—the Bering Sea between Russia and Alaska, the Okhotz Sea between the peninsula of Kamchatka and the main landmass of Asia, and the Timor Sea between Australia and Indonesia. And so this contribution to the slowing of the

Earth's spin rate depends greatly on ocean levels—which are related to the Earth's climate—and on the configuration of coastlines.

And there are other processes that have caused the spin rate to change. During the last ice age, thick ice sheets covered much of Canada and Scandinavia, the weight of those sheets causing the ground to subside, in places, more than a mile. The ice sheets melted much faster than the ground could recover, and so two of the larger depressions are still visible and are filled with water, forming Hudson Bay and the Baltic Sea. But the land continues to rise. As a result, just as the speed that a figure skater spins depends on whether the arms are stretched out or are held close to the body, the spin rate of the Earth also depends on how the planet's mass is distributed.

If all of this seems complicated, well, it is. It requires using Newton's theory of gravity not only to follow the Moon's motion and to understand the origin of tides, but also to know how the rotational momentum of the Earth's spin gets transferred to the Moon's orbital momentum, how tidal friction changes the Earth's spin rate, and how the Earth's interior recovers after an ice age. And there are other complications. For example, there seems to be a coupling between the spin of the Earth's fluid core and of its solid mantle, a process that transfers rotational momentum to different parts of the Earth, changing the Earth's spin rate. Overall, these complications can be quantified. And they can be used to explain current measurements that show the Moon is accelerating in its orbit—best expressed as a receding of the Moon away from the Earth at an average rate of about an inch a year—and that the Earth is spinning slightly slower so that the length of a day is increasing at a rate of about two milliseconds per century.

And those rates of the Moon's acceleration and the Earth's slowing spin rate, if used to calculate backward in time the Moon's motion, does show that a total solar eclipse *was* visible from Babylon on April 15, 136 B.C.E. In fact, a detailed history of the Earth's rotation rate has been determined, in large part, by using ancient eclipse records—though other records are not as complete or unambiguous as the one for 136 B.C.E.—and by using a trove of other celestial measurements, for example, timing the exact moment when the Moon passes in front of a star.

Here the astute reader should ask: Doesn't this affect the lists of ancient chronologies derived from ancient eclipse records and mentioned in an earlier chapter? Yes, it does. Both the Moon's gradual acceleration and the Earth's slower spin rate must be considered. And so the lists of ancient chronologies are constantly being adjusted. And more adjustments are anticipated as additional eclipse records are discovered and as the records that we now have are reinterpreted, and as a better history of the Moon's motion and the multitude of processes that affect its motion are better understood.

And this work is not solely of academic interest. It bears on other aspects of society. Notably, it pertains to questions about theology and the interpretation of scripture. To illustrate, for many years there has been a debate about the date of the crucifixion of Christ, that is, whether or not an *exact* date can be determined. The answer to that seems to rest on the timing of a lunar eclipse.

———————

The date of the crucifixion has long fascinated scholars because of its obvious importance in Christian theology and because it seems that there is some hope of answering it.

All who have addressed the question have agreed on a few basic facts. First, Pontius Pilate was procurator of Judaea at the time, which means the crucifixion must have occurred between the years of 26 and 36 C.E. Furthermore, given that Jesus was not baptized until at least 28 C.E., that he preached at least two years and that his death preceded the conversion of Paul, the range of possible years can be further reduced to between 30 and 34 C.E.

The four canonical gospels of Matthew, Mark, Luke, and John all state that the crucifixion occurred on the afternoon of a Jewish Sabbath, that is, on a Friday afternoon. This is confirmed in the writings of the Roman historian Tacitus. Furthermore, again according to the four gospels, the crucifixion occurred during the Jewish month of Nisan during the feast of Passover, which means during the spring and at the time of a full moon. One additional and highly important constraint comes from a statement by the prophet Joel mentioned in Acts 2:20 in which he said: "The moon will be turned into blood before the great and glorious day of the Lord coming." Many theologians have associated this statement to the crucifixion, suggesting that "the great and glorious day of the Lord coming" refers to the day when Jesus Christ died.

Given that association, the first part of Joel's statement takes on an important meaning. "The moon will be turned into blood" is a phrase often used to describe a total lunar eclipse, and that is the interpretation here. As the Moon moves into the Earth's shadow, it does not disappear completely. Instead, the disk takes on a dark reddish color, similar to that of blood, caused by the scattering of sunlight through the thin rim of the Earth's atmosphere that is visible from the Moon.

If one accepts the prophecy made by Joel as referring to Christ's death and that the crucifixion was concurrent with a lunar eclipse, then putting these together within the range of years already cited, and after extensive calculations, one concludes that the crucifixion of Jesus Christ must have occurred on Friday, April 3, 33 c.e., a date when a lunar eclipse was visible from Jerusalem.

Or was it?

In 1983 Colin Humphreys and Graeme Waddington of Oxford University did calculations that incorporated what was then the best estimate of the history of tidal dissipation and concluded that, when the Moon rose on Friday, April 3 in the year 33, as seen from Jerusalem, the Moon was in a partial lunar eclipse. In 1990 Bradley Schaefer, now at Yale University, did the calculations again, using what many considered to be a more realistic history of tidal dissipation and concluded that when the Moon rose on April 3, as seen from Jerusalem, a lunar eclipse had just ended, putting into doubt that this was the date of the crucifixion.

All of this is to show how crucial a knowledge of the history of the Moon's motion and the Earth's changing spin rate is when answering what seems to be a straightforward question. Whether the exact date of the crucifixion will ever be determined is still undecided. Perhaps a document will be discovered in the future that will clarify beyond all doubt whether a lunar eclipse actually occurred. It is also possible that the continued study of ancient eclipse records will further refine the history of the Earth's spin rate, leading to new calculations that do show that a lunar eclipse was visible from Jerusalem in the year 33. Either way, as the demanding work of searching for and compiling more evidence continues, and new interpretations are offered, one must keep in mind a point made by Humphreys and Waddington: Though ancient writers, such as those of the gospels, had many motives, they were not trying to provide obvious clues for future chronologists.

————◆————

The first attempt to predict the exact time when a solar eclipse would occur, and then to use a clock to time it accurately and check the prediction, was by Edmond Halley in 1715 during a total solar eclipse that was visible from near London. His predicted time was four minutes late, a respectable accomplishment considering that he had based his prediction on Newton's theory of gravity that had been proposed just a few decades earlier.

The next serious attempt came more than 125 years later, when in 1842 François Arago, secretary of the *Académie des sciences* and director of the Paris Observatory, did a vastly more complex set of calculations, taking into account the gravitational effects of Venus, Mars, and Jupiter to predict the time when a total solar eclipse would be visible from southern France. The difference between his predicted and measured times was a mere forty seconds. Ever since, those who planned eclipse expeditions have often included an accurate chronometer to time the event.

In 1869, from Burlington, Iowa, Maria Mitchell of Vassar College determined that the starting time of the total solar eclipse of that year was twenty-three seconds later than predicted. In 1878, from La Junta, Colorado, Asaph Hall of the United States Naval Observatory measured a time difference of twenty-nine seconds. An eclipse party that traveled to Spain in 1905 measured twenty seconds. A party sent to tiny Flint Island in the central Pacific Ocean in 1908 measured twenty-seven seconds. In 1918, William Campbell of Lick Observatory in California went north to Baker City, Oregon, and determined a time difference of fifteen seconds. Other eclipse observers in the late nineteenth and early twentieth centuries measured similar differences of fifteen to thirty seconds. And that seemed to be the limit, at least, with the current theory of lunar motion. A new mathematical strategy would have to be applied that used new ways to approximate the Moon's motion, one that would be achieved by two men who worked in succession on the problem for more than forty years.

A person who knew George Hill better than most described him as "modest to the verge of timidity." He never married. Those who did know him felt he preferred no personal contact with others whatsoever. He lived in Washington, D.C., and worked for the federal government in the office that prepared the annual *Nautical Almanac*. Though his income was small, he was known to refuse his salary, saying that he did not need the money and that it bothered him to look after it. But there were three things that he did delight in. He enjoyed long walks through the many forested areas that then surrounded the nation's capital. He also enjoyed being surrounded by books, especially those he kept at home as his own large library. And he was a mathematician, though his specific interest in mathematics was limited to what he needed to know to solve the problem of the Moon's motion.

An entire chapter would have to be devoted to describing the innovation he applied to understanding the Moon's motion. Let it be said that all previous workers had started with what is known in Newtonian theory as the two-body problem (the Earth-Moon system), and then determined what slight adjustments were needed by considering the gravitational attraction of the Sun, other planets, and lunar tides. Hill started with the much more complicated three-body problem (the Earth-Moon-Sun system) then added adjustments. In total, he incorporated in his calculations nearly 3,000 such adjustments. And those calculations involved manipulating numbers to fifteen decimal places. And all of this was done by hand, this being long before mechanical calculators. But there was not enough time for Hill to complete his new lunar theory. During his last years he collaborated with the man who would be his successor.

Ernest Brown was born in England and moved to the United States, becoming a professor at Yale University, to work with Hill on the problem of the Moon's motion. Brown was also a self-effacing loner. And he never married. He lived most of his adult life with his younger unmarried sister, Mildred. He was known to be a capable pianist, played chess at a high level, and loved to read detective stories. He had a particular interest in nonsensical verse, able to recite long passages from Gilbert and Sullivan operettas and from the writings of Lewis Carroll. But his major passion was the same as Hill's: to understand the Moon's motion.

Brown completed what is now known as the Hill-Brown theory in 1908. He then took eleven years to apply it, computing tables of the Moon's position up to the year 2000. A major test of the theory came in 1925 during the total solar eclipse visible from the northeastern United States, including the northern half of New York City. Those who accurately timed the eclipse reported that the predictions made using the Hill-Brown theory were within *one second* of the observed times.

It was a remarkable achievement, showing how the Moon moves with mathematical precision across the heavens. For the next several decades, the Hill-Brown theory was the standard used to predict the Moon's motion. In the early days of rocketry, it was used to plan the trajectories of the spacecraft sent to the Moon. Since then, the Hill-Brown theory has been replaced by a succession of theories.

The latest is known by the innocuous name JPL DE406*, a math-ematical procedure that accounts for the gravitational effect of every major body in the solar system, including many planetary moons and several large asteroids. And using JPL DE406, the difference between the predicted time of an eclipse and the time when it actu-ally happens has been reduced by half. That is, the exact moment when the Moon will completely obscure the Sun, computed several years in advance, is now known to an accuracy of about one second.

And that is where it has stood for the last few decades. It has not been possible to reduce further the one-second accuracy. And we know why. As mentioned at the beginning of the chapter, the Earth does not spin at a steady rate. It is slowing down, mainly due to the transfer of the Earth's rotational energy to the Moon's orbital energy through ocean tides. There is also a component due to the rebound of the Earth's surface from the melting of ice sheets since the latest Ice Age. These we understand, and we can account for them. But there are smaller effects that cause the length of a day to change a millisecond or more over the course of a year in a haphazard way that we cannot account for. These may be due to extreme weather, such as a heavy snowpack in one region, or the occurrence of several large earthquakes. Even changes in the migra-tory patterns of birds have been suggested as a cause. The bottom line is this: The spin rate of the Earth is changing by a very small amount in ways that cannot yet be predicted. It is this uncertainty that limits our ability to predict, years in advance, the timing of an eclipse to about one second.

Extensive use has been made of the current theory of lunar motion to produce massive catalogues of eclipses, both of the past and

---

* JPL DE406 stands for the Jet Propulsion Laboratory Developmental Ephem-eris No. 406, which was developed at the Jet Propulsion Laboratory in Pasa-dena, California.

for the future. Because historical records of eclipses go back only a few thousand years and because the spin rate of the Earth can be reasonably estimated for only about a thousand years into the future, such catalogues are usually confined to a 5,000-year time span, from about 2000 B.C.E. to about 3000 C.E. And there is much that can be learned from them.

For example, one of the basic questions about a total solar eclipse is: How often can one be seen from any given spot on Earth? On average, in the northern hemisphere, one will be seen every 330 years and, in the southern hemisphere, about every 380 years. The difference is caused by the slight eccentricity of the Earth's orbit, that is, the Earth's path around the Sun is approximately an ellipse, and so the Earth-to-Sun distance varies slightly during a year. During the summer months in the northern hemisphere the Earth is about three million miles farther from the Sun than during the winter months. And the reverse is true in the southern hemisphere. The Sun is also above the horizon longer during the summer months. These factors—the Sun being farther away (and so it appears smaller in the sky), and the daytime being longer during summer months—account for the difference.

But those are average intervals. Because eclipse paths crisscross the Earth's surface, there are places where those paths must intersect. One such coincidence happened recently along the Atlantic coast of Africa where two total solar eclipses were seen from the country of Angola just eighteen months apart, in June 2001 and again in December 2002.

There have also been triple intersections of eclipse paths, though such intersections happen over periods of decades. If one lived in the small mining town of Telegraph Creek in the Yukon Territory of Canada in the nineteenth century, three total solar eclipses would have been seen in just twenty-seven years, in 1851, 1869, and 1878. If one wanted to identify an eclipse capital within the United States it would be Baker City, Oregon, where, in less than a hundred years, three total solar eclipses were or will be visible, in 1918, 1979, and 2017. There will be another notable triple intersection, though just outside the United States, that will occur in northeast Mexico just south of Brownsville, Texas, in 2052, 2071, and 2078.

And then there is the opposite of a quick succession of total solar eclipses, that is, there are eclipse droughts. In ancient times, no one living in the city of Jerusalem would have seen a total solar eclipse between 1131 and 336 B.C.E.—an eclipse drought of 795 years—but, then, after 336 B.C.E., there were three total solar eclipses visible from the city in just fifty-four years.

Within the United States, an eclipse drought is about to end for the small community of North Platte, Nebraska. The most recent total solar eclipse visible from that location was on July 29, 957 C.E. The next will be on August 21, 2017, an eclipse drought of 1,160 years.

An eclipse drought lasting more than six hundred years is rare. But the complex motions of the Sun and the Moon and the changing spin rate of the Earth have conspired so that there are *three* places in the state of Ohio that are currently experiencing such a drought. The last total solar eclipse visible from Columbus, Ohio, was in the year 1395. The next one will be in 2099, a period of 704 years. Seventy miles west of Columbus is Dayton, Ohio, where the current eclipse drought began in the year 831 and will end 1,192 years later on April 8, 2024. And in Cincinnati, Ohio, south of Dayton, the most recent total solar eclipse happened on January 21, 1395. The next one will be visible sometime after the year 3000, an eclipse drought that will have lasted more than 1,600 years.

––––––•––––––

Once it has been established where and when one should be to see a total solar eclipse, the next challenge is a dual one: How to guarantee a cloud-free sky and how to maximize the duration of darkness.

The obvious answer to the first is to find a way to be above any potential clouds. In that regard, the first attempt to do so was made in 1887 by the Russian chemist and originator of the periodic table of chemical elements, Dmitri Mendeleev.

For many years, Mendeleev had pursued an interest in the behavior of gases. That led to an interest in ballooning. And so, when he was given a chance to witness a total solar eclipse from a balloon, he took it.

The path of totality was to pass directly over the village of Klin several miles northwest of Moscow. That was where Mendeleev decided to make his ascent. With the support of the Russian Imperial Institute of Technology, a hydrogen-filled balloon was provided, as was a pilot.

The weather was wet the night before the eclipse, so that by morning the balloon was covered with heavy dew. Mendeleev and the pilot climbed into a basket on the underside. An attempt was made to cast the balloon loose, but it refused to ascend. By one account, the pilot decided to climb out and make some adjustments, but, relieved of his weight, the balloon suddenly rose, and amid the jostling and hand clapping of the crowd, the pilot was unable to return and Mendeleev was carried upward. As he rose higher into the air, it was said that he called down to several friends and asked them to promise to collect his bones.

The balloon carried him up through a low blanket of clouds where he had an excellent view of the eclipse, though few people on the ground were able to see it. He also saw the Moon's shadow pass over the top of the clouds. Then for the next hour he saw nothing as the balloon slowly descended, landing him on the grounds of a monastery more than a hundred miles away. Thus ended the first attempt to do airborne astronomy.

Other attempts were made to observe solar eclipses using balloons or other types of lighter-than-air aircraft. The most ambitious attempt was made in 1925 with the Navy dirigible *Los Angeles* that carried aloft a dozen scientists and a crew of nearly thirty to observe a total solar eclipse from the eastern tip of Long Island. But these were soon supplanted by a more versatile invention to reach a high altitude—the airplane.

Less than a decade after the Wright brothers made the first powered flight in a heavier-than-air aircraft, a solar eclipse was first observed from an airplane. It was April 17, 1912, and the event was an annular eclipse. Two men, Michel Mahieu, already an accomplished pilot at the age of 20, and Gaston de Manthé, his navigator, took off in a military biplane from a field near Paris. They flew close to the Eiffel Tower, then westward several miles to Saint-Germain-En-Laye where from an altitude of about a thousand feet they saw the Sun almost covered completely by the Moon. The next day the newspaper *Le Matin* paid homage to the two men who had been "enchanted by having been able to observe the eclipse from a little closer than ordinary mortals."

David Todd, the professor of astronomy at Amherst College and an early aerial enthusiast, tried twice to be the first person to observe a total solar eclipse from an airplane. His first attempt was in August 1914 when he and a French pilot went to Russia, but Russian authorities prevented them from flying because the First World War had just begun. Todd made a second attempt in 1919 when the United States Navy provided him with one of its first seaplanes and six of its personnel—two pilots and four machinists—to support his effort. The plan was for Todd and one of the pilots to fly the seaplane to an altitude of about 10,000 feet. Then, when the Moon's shadow was over them, for the pilot to put the seaplane into a sharp dive, increasing its speed, so that the duration of totality would be increased by a few seconds. The flight was to take place from Montevideo in Uruguay. But a storm hit the night before the eclipse and the seaplane was destroyed.

Another attempt to see a total solar eclipse from airplanes came on September 10, 1923, in California, when the United States Navy sent sixteen planes aloft, each with a pilot and a photographer, when a path of totality passed close to the Naval Air Station at San Diego. Unfortunately, the sky was overcast, the clouds too high for airplanes of the era to fly above, and as the official report read: "No valuable results were secured." In 1925, yet another attempt was made, this time by twenty-five airplanes flying from the Army base at Mitchel Airfield on Long Island. The sky was clear. But the vibrations inside the airplanes were too severe, and no clear photographs of the corona were made.

Finally, a minor feat in aviation history was achieved on August 31, 1932, when the corona was successfully photographed from an airplane. On that day the path of totality crossed over Maine and New Hampshire. The photographer was Captain Albert Stevens of the United States Army Air Corps, a noted aeronaut who sat with his equipment in the rear seat of an open-cockpit biplane. He and the pilot flew to an altitude of 27,000 feet. Stevens communicated to the pilot by yelling short shrieks, or "yips." One yip meant turn left, and two quick yips meant turn right. In addition to the corona, he also managed to photograph the Moon's shadow as it approached.

The speed of the airplane flown in 1932 increased the duration of totality by two seconds. The first significant increase occurred in 1955 when a Lockheed T-33 jet, flying at nearly six hundred miles per hour, almost doubled the duration of totality from about seven minutes to more than twelve minutes. But the most impressive extension of totality occurred in 1973 when the supersonic transport, the Concorde, was flown under the Moon's shadow at Mach 2.08.*

The Concorde took off from the airport on the island of La Palma in the Canary Islands just as the Moon started to move in front of the Sun. Forty-five minutes later, after flying more than four hundred miles, it was in the shadow. From there, it flew at 56,000 feet, traveling 1,800 miles across Africa along the eclipse path, extending the duration of totality—which was a maximum of seven minutes four seconds on the ground—to seventy-four minutes!

Five holes had been cut into the fuselage of the Concorde as viewing ports for scientific equipment; each hole covered with a clear quartz window. The idea was to take advantage of the extra long duration of totality to record changes in the corona. In one particular experiment, it was shown for the first time that the brightness of the corona does change over a period of about five minutes. In another

* This is 1,340 miles per hour, more than twice the speed of a commercial jet airliner.

experiment, which trained a camera on the sky far from the Sun, something unusual was seen. It was a suspicious dash of light just above the horizon. When the film was checked, it was determined that this was not a defect or scratch on the film. It was white in the middle, orange-red on top and fringed in green. Might it be a UFO? Might it be extraterrestrials interested in the cosmic scene that was taking place in the sky, as well as fascinated by the fast-flying machine?

The French space research institute, *Centre national d'études spatiales* (CNES), which has a special department to investigate unidentified objects spotted in the sky, gave the photograph a thorough examination. Finally, six months after the eclipse, CNES released its conclusion. The object was a small cloud that had condensed when a meteor, about the size of a grain of sand, had passed through the upper atmosphere, something that is not unusual. The orange-red color was attributed to unusual lighting caused by the eclipse. The green fringe, similar in color to the aurora borealis, was attributed to oxygen atoms suddenly stimulated to a higher energy state by the passing of the fast-moving object. This is an all-but-forgotten episode in the history of eclipse watching, except by UFO enthusiasts who continue to mention it, even today.

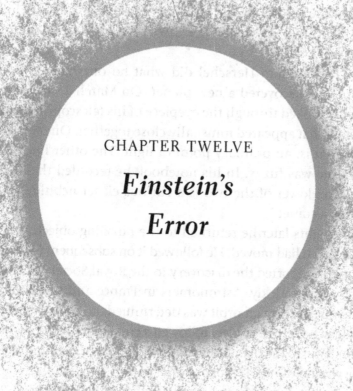

CHAPTER TWELVE

Einstein's
Error

A dark, dark, dark, amid the blaze of noon,
Irrecoverably dark, total Eclipse
Without all hope of day!

—John Milton, *Samson Agonistes*, 1671

To say that Newton's theory of gravity was at the center of all work done between the seventeenth and nineteenth centuries about the motions of the Moon and the planets is an understatement. It completely dominated them. Whenever a discrepancy arose, no matter how slight, between a predicted position and an actual one of the Moon or a planet, the theory could account for it by the introduction of a new mass. This was most strikingly shown by the discovery in 1846 of the planet Neptune.

In 1781 William Herschel did what no one had yet done in history. He discovered a new planet. On March 13 of that year, Herschel looked through the eyepiece of his telescope to examine two stars that appeared unusually close together. One was clearly a star, that is, an ordinary point of light. The other looked odd. Its outline was fuzzy. In his notebook he recorded the observation: "The lower of the two is a curious either nebulous star or perhaps a comet."

Four nights later he returned to the puzzling object, this time noting that it had moved. He followed it on subsequent nights, and, eventually, reported the discovery to the Royal Society of London. Word spread quickly. Astronomers in France and Germany took up the quest. Soon an orbit was determined, an orbit that showed that Herschel's fuzzy object followed a circular path far beyond the one for Saturn, the planet then known to be farthest from the Sun. There was no doubt. It must be a planet.

The newly discovered object was given the name *Uranus*, after the ancient Greek god of the sky. And astronomers kept following it across the night sky, noticing over the next few decades that its movement did not quite correspond to that predicted by Newton's theory of gravity. Until the 1820s, the new planet was always slightly ahead of the predicted position, then afterward, it started to fall behind. The reason seemed clear. There must be yet *another* undiscovered planet orbiting at a distance beyond that of Uranus, its gravitational attraction responsible for perturbing the movement of Uranus.

Among those who took up the quest to discover where this other planet might be hidden was French astronomer and mathematician Urbain Le Verrier. But he did not head to a telescope. Instead, he spent months engaged in complex calculations to explain the small but systematic discrepancies between Uranus's observed motion and the one predicted by Newton's theory of gravity. On September 18, 1846, having completed his work, Le Verrier sent a letter to a German astronomer, Johann Galle at the Berlin Observatory, predicting where a new planet beyond Uranus would be found.

The letter arrived five days later on September 23. That evening Galle positioned his telescope on the spot on the sky given by Le Verrier. After only ten minutes of searching, studying every object in view, Galle realized that one showed a fuzzy outline. He checked his star chart. The fuzzy object was not on it. He was looking at a new planet, one that would be named *Neptune*.

It was a further confirmation of Newton's theory. And it made Le Verrier a hero. He received worldwide acclaim for "using his pen" to discover an unknown planet. He received the *Légion d'honneur* from Louis Philippe, the king of France. And the French government promised to provide funds for his future work. But what would he do next?

Even before he had started looking into the discrepancies in the orbit of Uranus, Le Verrier had noticed a slight anomaly in the orbit of Mercury, the planet closest to the Sun. According to Newton, if only one planet orbited around the Sun, that planet would follow exactly the same path every time it circled the Sun, a path that would be an ellipse with the Sun at one focus. However, because other planets were present, the orbit of each one was perturbed slightly from an ellipse. In the case of Mercury, the gravitational attraction of the planets caused Mercury to follow a spiral pattern in such a way that the point where Mercury made its closest approach to the Sun, known as the perihelion, shifted slightly with each orbit. The slight shift, known as precession, had been measured to be 565 arc-seconds per century.* After Le Verrier considered the gravitational influence

---

* An arc-second is an angular measurement. A circle has 360 degrees. An angle of one degree is composed of 60 arc-minutes, and an angle of one arc-minute is composed of 60 arc-seconds. That is, one arc-second is 1/3600th of a degree.

of all the planets, he could account for a rate of 527 arc-seconds per century. The difference, 38 arc-seconds per century, though small, was not negligible. And it needed to be explained.[*]

In 1859 Le Verrier proposed a solution. A planet, yet unseen, might be orbiting very close to the Sun, closer than Mercury's orbit, and its gravitational influence might account for the small anomaly in Mercury's motion. Le Verrier did the calculation, predicting that the yet-to-be-seen planet might lay halfway between Mercury and the Sun and be roughly the same size as Mercury. But, if that was the case, why hadn't anyone seen it?

Someone apparently had—and recently. A certain Dr. Edmond Lescarbault who practiced in the little country town of Orgères-en-Beauce several miles south of Paris had been studying the Sun through his telescope—something that he regularly did—and noticed, on March 26, 1859, that a dark object moved across the disk of the Sun. He had been unable to watch the entire passage, presumably because the needs of patients intervened, but he was sure, after learning about Le Verrier's prediction, that he had indeed seen the object. And so he sent a letter to Le Verrier in Paris announcing the discovery.

At first Le Verrier was skeptical, but he went to see Lescarbault anyway on the chance that maybe he had seen something. He checked the doctor's equipment. Lescarbault had constructed a modest observatory at one end of a stone barn. Inside he mounted what, to Le Verrier, looked like a first-rate telescope. They discussed Lescarbault's observation. Lescarbault had not kept a written record, but he did return to the telescope several times that day, and most assuredly had seen something cross in front of the Sun. And he was almost sure that it had taken almost an hour.

---

[*] To be historically accurate, in 1882 Simon Newcomb of the United States Nautical Almanac Office recomputed the gravitational attraction of the other planets on Mercury using better values for the masses of each planet than Le Verrier had. According to Newcomb's calculation, the precession rate of Mercury's perihelion that could not be explained by Newtonian gravity was 43 arc-seconds per century, the rate that was later used by Einstein.

Le Verrier was convinced. On January 2, 1860, he announced at a meeting of the *Académie des sciences* in Paris that, in his opinion, Lescarbault had seen a new planet, which, because of its nearness to the Sun, Le Verrier named *Vulcan*. The news hit the popular press. And with that came a flurry of reports from people who claimed to have already seen the planet.

A Mr. Benjamin Scott of London wrote to the *Times* to say that he should be credited with the discovery, having seen Vulcan "at or about Midsummer 1847." Rudolf Wolf, an astronomer living in Zurich and a keen observer of sunspots, provided a list of twenty-one observations he had made over the previous few decades, some as early as 1819, that he said were probably the newly discovered planet moving across the Sun's disk. A report also came from Emmanuel Liais, a French astronomer, botanist, and explorer who was in Brazil at the time. Liais, too, had pointed his telescope at the Sun on the day Lescarbault made his historical discovery and, according to Liais, he had seen nothing but normal sunspots. And so he questioned Lescarbault's discovery and criticized Le Verrier for accepting it without stronger evidence.

Nevertheless, excitement about the discovery continued to build, though it was widely acknowledged that it had to be confirmed. A clear sighting by several observers had to be made. And the most opportune time for that to happen was during a total solar eclipse.

Le Verrier traveled to Spain to search for Vulcan during the July 18, 1860, total solar eclipse and saw nothing. He went to Thailand for the same purpose in 1868, and again nothing, even though the period of totality was unusually long, lasting for more than six minutes. The next opportunity was also a good one.

It was over the United States in 1869. And dozens of American astronomers were out in force, many of them including a search for Vulcan in their plans. No one saw it. Benjamin Gould of Harvard College made a special attempt using four different cameras to photograph Vulcan. He observed the eclipse from Burlington, Iowa, where the sky was clear. He took forty-two photographs and studied them carefully. None showed the elusive planet. He collected photographs made by other astronomers, a total of four hundred additional images. He examined them. Stars were clearly visible in them. But there was no sign of Vulcan.

The interest in Vulcan now waned and it probably would have ended when Le Verrier died in 1877, but, a year before his death, a report suddenly came from China that the elusive planet had again been spotted.

The report came from a Mr. Weber who informed astronomers in Europe of what he had seen. He described it as "a small well-rounded disk" passing in front of the Sun. He also mentioned that, just before his observation, the sky had cleared suddenly.

Weber made his observation on April 4, 1876. After that, there was another rush of reports by people who were claiming to see Vulcan. A person identified only by the initials "B.B." from Montclair, New Jersey, wrote to the editors of *Scientific American* to report that he saw a round spot move across the Sun on July 23. W. G. Wright of San Bernardino, California, after reading B.B.'s letter, aimed his telescope at the Sun and immediately spotted a moving object. Samuel Wilde, also of Montclair, wrote to the magazine to report his sighting, as did a Mr. John H. Tice of St. Louis, Missouri. So many reports were being received that, at the end of the year, the editors of *Scientific American* informed readers that they would cease to print any more reported sightings of Vulcan.

And yet, with the dozens of new reports, none of them had been made simultaneously, that is, no two observers had independently seen Vulcan at the same time. And that was reason to doubt whether the planet did exist. But there would soon be another excellent opportunity. The Sun would again soon be eclipsed, and it would again happen over the United States where scores of professional and highly experienced amateur astronomers would gather along the narrow path of totality. Surely, several of them would see the long-sought planet of Vulcan.

James Watson of the University of Michigan at Ann Arbor was espe-cially keen to find the planet. He was completely convinced of Le Verrier's calculations, Lescarbault's initial observation, and all of those that followed. Vulcan had to exist. It was a matter of having an experienced astronomer positioned at the right place and the right time to see it. And Watson had the profession credentials. He was in charge of the Detroit Observatory at the University of Michigan and had used that telescope to discover, by naked-eye observation, twenty asteroids. He had also been on eclipse expeditions in 1869 to Burlington, Iowa, and in 1870 to Sicily. The 1878 eclipse over the western United States would be his third. And he had chosen Wyoming to view the eclipse, expecting to be the first of many who would sight the planet.

Watson set up his eclipse station near Separation, Wyoming, about a hundred yards from where Simon Newcomb of the Nautical Almanac Office had set up his. The two would coordinate their obser-vations. Newcomb would begin by looking for solar prominences and studying the shape of the corona. Watson would devote himself entirely to searching for Vulcan. If he saw a candidate, he was to rush to where Newcomb stood and he would look for the planet and con-firm its existence. Then Watson would return to his telescope, and reconfirm his observation. As soon as the eclipse was over, Watson would write out a brief description of where Vulcan was located, hand the description to a man awaiting on horseback, ready to gallop a quarter-mile to the nearest telegraph station, and have the message sent to Dallas, Texas, where it would be delivered to David Todd of Amherst College who would then have several minutes before the Moon's shadow passed over him to plan where to look for Vulcan.

A half hour before the eclipse was to begin, Newcomb walked into a darkroom where photographs were to be developed and sat in the darkness adjusting his eyes, making them as sensitive as pos-sible. Three minutes before the eclipse, he emerged. Watson was already at his telescope.

The Moon's shadow came over them. Both men proceeded to make their observations. Watson looked carefully. He had a star chart and he recognized a few stars as he looked through the telescope. But one of the star-like objects was *not* on the chart. He searched some more. He saw a *second* object that was not on the chart. There must be *two* undiscovered planets orbiting the Sun! He ran to where Newcomb was working. A steady wind had caused a delay in his observations. With a few seconds left, he began to look for Vulcan where Watson had instructed. Meanwhile, Watson raced back to his telescope to reconfirm the sightings. By then, totality was over. And for some reason, Watson never wrote out his description of where he had seen the two possible planets. Newcomb was not able to confirm the sightings, nor could Todd. But to his dying day—he lived just two more years—Watson maintained that he had seen *two* objects that must be planets in close orbit around the Sun. He even gave a name to the second one. He called it *Adonis*.*

Only one other person claimed also to see a Vulcan-type planet close to the Sun. He was Lewis Swift, an amateur astronomer from Rochester, New York, who observed the 1878 eclipse from Denver. Like Watson, Swift was also an experienced observer, having discovered several comets. However, the star-like planet that he saw in 1878 was not close to either of the positions of the two objects Watson had seen. And so the 1878 solar eclipse came and went, and there was still no confirmed sighting of Vulcan.

Astronomers continued to make attempts to see Vulcan during solar eclipses, though with slightly less enthusiasm. A total solar eclipse over central California in 1880 yielded no sightings. One over Egypt in 1882 when totality was brief, lasting less than a minute, also yielded no planet, though a comet near the Sun was seen clearly. Similar results came from solar eclipses in 1889, 1898, and 1901. In 1905 astronomers from Lick Observatory in California organized three eclipse parties, sending them to Canada, Spain, and Egypt. The observers in Canada and Spain saw only clouds.

---

* At the time of his death in 1880, Watson was building a special observatory to look for Vulcan and Adonis, still believing that both planets existed.

Those in Egypt took photographs that showed the images of fifty-five individual stars, but no unknown planet.

Three years later, in 1908, astronomers from Lick Observatory sent an expedition to tiny Flint Island in the central Pacific Ocean, one of only two islands where the path of totality would cross. Astronomers from the Royal Astronomical Society of Canada joined them. The astronomers arrived a month before the day of the eclipse. Among their equipment were four specially designed cameras to be used to search for Vulcan. Each one had a three-inch lens and an eleven-foot focal length.

On the day of the eclipse, January 3, a heavy rain was falling and all seemed to be lost. Miraculously, a minute before the beginning of totality, the rain suddenly stopped and the sky cleared. The astronomers did their work. A single photographic plate was exposed in each of the special cameras for three minutes. After development, the plates were examined. More than 300 stars could be identified. Again, no planet Vulcan was seen.

From the sensitivity of the plates, it was estimated that any object larger than thirty miles in diameter orbiting the Sun should have been detected. To further diminish the case for Vulcan, more than a million such planets of that size would have to be present to account, by gravitational perturbation, for the motion of the perihelion of Mercury.

That negative result effectively ended the search for an intramercurial planet. It also caused many scientists to consider a radical alternative: Might it be necessary to modify Newton's theory of gravity to account for Mercury's peculiar motion?

————•————

In 1905, while working as a patent clerk in Bern, Switzerland, Albert Einstein published his special theory of relativity. It was "special" in that it applied to a very specific type of motion, that is, motion at a constant speed, vis-á-vis the speed of light. In this special theory, there was no acceleration. In nature, however, almost everything seems to be accelerating: When one is jostled in a car, when one climbs a staircase or slides down a banister, or when the Earth rotates on its axis. It also said nothing about gravity, which meant it could not account for the motion of planets as they orbited the Sun. Such shortcomings were an embarrassment. And they needed to be addressed.

His next major advance came at the end of 1907 while he was writing a summary about special relativity for a science yearbook. It was at that time that he conceived a thought experiment in which he wondered what one would see and feel if one was inside a windowless box and the box was either falling freely or accelerating.

For example, imagine yourself inside an elevator and the cable holding the elevator breaks. The elevator now starts to fall. Because everything falls at the same rate—a conclusion that dates back to Galileo in the seventeen century—you and everything else inside the elevator will be weightless. You now take a flashlight and shine a light on the elevator wall. Because the speed of light is finite, it takes a finite amount of time for the beam of light to reach the wall. And if you look at the path the beam of light takes, it will be a straight line. Now imagine you are inside the same elevator, but are now in outer space. Again, you and everything inside the elevator are weightless. But, this time, a rocket engine has been attached to the bottom of the elevator. When the engine is turned on, the elevator is now accelerated. What happens inside the elevator? You and everything inside the elevator would drop to the floor. If you shine a beam of light onto a wall, because it takes a finite amount of time for the beam to reach the wall, the beam will curve down toward the floor. Now comes the great insight that Einstein had.

If, instead of being inside an elevator in outer space accelerated by a rocket engine, you are standing inside the same elevator but on the Earth's surface, everything would be the same. That is, the laws of physics are the same whether you and everything around you are being accelerated or whether you and everything around you are in a gravitational field. And that means, just as a beam of light curves when the surroundings are accelerated, a beam of light will also curve when passing through a gravitational field.

It was another four years before Einstein did the calculation. The largest gravitational field in the solar system is of course due to the Sun. And so he calculated how much a beam of light from a distant star would bend if it just grazed the Sun's rim. He determined the deflection would be 0.87 arc-seconds. He also suggested when such a deflection might be measured: during a total solar eclipse.

Two photographs are taken of the same star field, one during a total solar eclipse when stars are often visible and one at night without the Sun present, that is, a few months before or after the eclipse. If the two photos are compared, the positions of the stars will have changed slightly. That is, during the eclipse, the positions of the stars would have seemed to move away from the Sun. He proposed this in a paper he sent to the *Annalen der Physik* in June 1911, ending with the statement: "It would be a most desirable thing if astronomers would take up the question." And they did.

Campbell of Lick Observatory was one who realized the importance of such a question, that is to determine if starlight was deflected by the Sun's gravitational field. And he had the equipment to do it: the cameras designed to discover Vulcan were ideal for the search. And he realized there would soon be an opportunity: a total solar eclipse over the eastern half of Europe in 1914.

Campbell decided on a site at the city of Brovary, near Kiev, close to the centerline of totality and serviced by railroads. Campbell and the rest of the eclipse party and the equipment arrived on July 21. The eclipse would be a month later on August 21. Between those two dates, on August 1, Germany declared war on Russia, beginning the First World War.

The members of a German eclipse party, led by Erwin Freundlich of the Berlin Observatory, were south of Campbell in the Crimea and were captured by the Russian army and their equipment confiscated. They were later exchanged for Russians who had been arrested in Germany. David and Mabel Todd of Amherst were near the city of Kamenka about a hundred miles southeast of Kiev. He had planned to use an airplane with a French pilot to photograph the eclipse, but the Russians confiscated the airplane because of the war.

Campbell and his party of Americans, which included his wife, three sons, and his mother-in-law, were not arrested by the Russians after the seizure of their plane, however, and were allowed to continue to prepare for the eclipse. Days in advance, the four Vulcan cameras and other equipment were arranged and tested. But the eclipse was not seen, and no photographs were taken. As Mrs. Campbell recorded it in her diary: "Total failure. Thick gray clouds at eclipse time and lovely clear sunshine afterward."

Afterward, the equipment was confiscated, and though nonmilitary freight was not officially being carried by the railroads, a Russian general intervened and Wallace's eclipse equipment was sent to the Russian National Observatory at Pulkovo where it would be stored until it could be sent back to the United States. Wallace and his party had originally planned to return to the United States through Berlin. That was now impossible. And so, again, the Russian general intervened and arranged safe passage for them to Petrograd. From there, after numerous delays, they managed to travel through Finland, Sweden, and Norway, then to London and finally back to the United States.

It was with a heavy heart that Campbell wrote to a friend about the disappointment he felt over his trip: "If the critical two minutes had been clear so that we could have brought valuable results home with us we would have thought nothing of the inconvenience which later greeted us at various points." This was his fifth eclipse expedition, and the first one that had failed. Never before, he would write, had he known "how keenly an eclipse astronomer feels his disappointment through clouds." He ended the note by writing: "One wishes that he could come home by the back door and see nobody."

And Campbell was not the only person disappointed. Einstein had made a prediction about the bending of starlight that had yet to be confirmed. But, in that regard, the failure of the 1914 eclipse expedition was a blessing in disguise because it provided an important twist to history. By now, Einstein realized he had made a mistake.

The next year, 1915, after mastering a novel branch of mathematics known as Riemannian geometry, he was able to propose a set of equations to explain how mass and gravity interact. It was the theory of general relativity.

The basic idea of general relativity is simple. Newton considered time and space to be absolute and distinct. Einstein saw them as connected through something that is now known as the space-time continuum. Furthermore, according to Einstein, the space-time continuum is not flat, in that it is not like the top of a pool table. Instead, it is warped and has hills and valleys, much as the elastic sheet of a trampoline gets warped when several people stand on it. To Newton, gravity is a force that causes objects to follow curved paths. To Einstein, those same curved paths are objects that are rolling along the hills and valleys of the space-time continuum.

To be more specific, to Newton, a planet follows a curved path around the Sun because the gravitational force of the Sun continues to pull on it. To Einstein, a planet follows a curved path because the space-time continuum is warped in a certain way. Again, using the analogy of a trampoline, if a heavy object, such as a bowling ball, is placed on it, the trampoline—our space-time continuum—sags. If you now roll a marble across the trampoline, the path it follows will curve because of the sagging.

Throughout the solar system, the difference between Newton's theory and Einstein's theory is very small, though the difference is larger for objects that are closer to the Sun. That immediately brought Einstein's attention to Mercury and its anomalous motion.

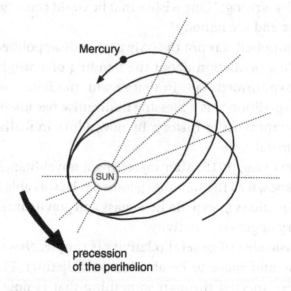

Mercury

SUN

precession
of the perihelion

Mercury's orbit around the Sun is that of a spiral, so that the point of its closest approach, the perihelion, precesses around the Sun.

He calculated what the precession rate of Mercury's perihelion would be using the theory of general relativity. He determined a rate of forty-three arc-seconds per century, the same as that observed by astronomers and unexplainable by Newton's theory. "I was beside myself with joyous excitement," Einstein recalled feeling after completing the calculation. "The results of Mercury's perihelion movement fills me with great satisfaction." He then moved on to determine how much starlight would be deflected by the Sun.

It must have come as a surprise, now that he had the full equations of general relativity to work with, that a beam of starlight that grazed the Sun's limb should be deflected 1.74 arc-seconds, twice the original estimate. If Campbell had succeeded in photographing the 1914 eclipse and had measured that amount, then it would have seemed that Einstein had jury-rigged his theory later to match the observed value. And Einstein might not have reached the worldwide acclaim that he did. But that is not what happened.

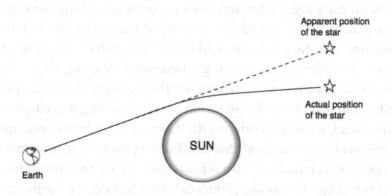

Deflection of starlight caused by the Sun's gravitational field.

Instead, on November 25, 1915, Einstein gave a presentation at a meeting of the Prussian Academy of Sciences in which he announced that general relativity could explain the precession rate of Mercury's perihelion and that the deflection of starlight as a result of this theory should be 1.74 arc-seconds. But because of the First World War, his announcement was not widely known outside of the German-speaking world. One person who did learn of it was Willem de Sitter of Leiden University in The Netherlands who, because of his country's neutrality, had access to Einstein's work.

De Sitter soon received a paper published by Einstein that described the prediction. It was filled with complex mathematics that few mathematicians yet understood, including de Sitter. But he knew someone who might be able to comprehend it. And so, though Germany and England were at war, and their citizens, including their scientists, were accusing each other of immoral acts and of abetting a level of death and destruction hitherto unknown to history, de Sitter forwarded the recent work of a German physicist to a British astronomer—Arthur Eddington of Cambridge University.

From an early age, Eddington liked to count up to large numbers. As a youth, he attempted to count all of the letters in the Bible, reporting that he got to the end of Genesis before giving up. To those who were with him at night beneath a clear sky, they could always be assured that, whenever the conversation dropped, Eddington would turn his attention upward and begin to count the stars. This fascination with counting naturally led him to manipulating numbers, and from that to a lifelong study of mathematics. At age nineteen he received a scholarship to study mathematics at Trinity College, Cambridge. Four years later he took the highly competitive "Tripos" exam, receiving the honor of "Senior Wrangler," meaning first place. After graduation in 1913, he was appointed a professor at Cambridge, earning the reputation as both the worst and the best lecturer. He was considered the worst because he seemed to begin every lecture in mid-sentence and did not stop until the end of the hour. He was the best because, as students would later attest, he talked a lot about big issues and addressed big questions and, though his train of thought was often difficult to follow, his major points were long remembered after a lecture was over. And there was another element that was important and guided him throughout his life. He was a Quaker and was dedicated to nonviolence.

In 1916, as the number of war casualties continued to rise, the British government imposed compulsory military service on all unmarried men between the ages of eighteen and forty-one. Eddington was then thirty-four, and though he was nearsighted and might not have to serve, was prepared to claim an exemption as a conscientious objector. But before that happened, the astronomer royal, Frank Dyson, who always had the ear of British military leaders, intervened and argued that as a scientist Eddington could best serve his country by continuing his professional work. The tribunal that was to hear Eddington's appeal agreed. And so he stayed at Cambridge University.

But by the spring of 1918 the war had worsened for the British. A German offensive was advancing. And more men were needed for the army. The tribunal notified Eddington that his exemption had been overturned and that he was to report for military duty. This time Eddington presented himself to the tribunal and claimed a new exemption based on his religion as a Quaker. That was rejected. Dyson again intervened. He wrote a letter to the tribunal informing the members that one thousand pounds sterling had been given to him by the British government to make preparations to observe a total solar eclipse in May of the next year "on account of its exceptional importance." He also noted that Professor Eddington was "peculiarly qualified to make these observations" and requested that the tribunal give Eddington permission to undertake the task.

When Eddington appeared again before the tribunal, in July 1918, its members wanted to know more about the eclipse. He explained how the total solar eclipse on May 29, 1919, would be the best opportunity for many years to test Einstein's theory, because on that date the Sun would be situated in front of a bright field of stars, the Hyades, perfect for measuring the deflection of starlight. It would also be an unusually long solar eclipse, lasting nearly seven minutes. Without retiring to discuss the case, the members decided to grant him an exemption from military service for an additional twelve months on the condition that he confine himself to preparing for and observing the eclipse.

He started preparations immediately, but because the war was continuing he had trouble finding skilled workers who could spare time to make the required, highly precise equipment. Fortunately, the tide of the war soon changed in favor of the British, leading to an armistice on November 11. After that date, the eclipse work proceeded in earnest.

Unlike other eclipse expeditions that planned to do many different types of observations, Eddington's expedition would focus on a single objective: to photograph stars near the Sun. He organized two parties for his expedition, sending them to different places to reduce the chance of being wiped out by bad weather. He would lead the party that would go to the island of Príncipe off the coast of West Africa. His counterpart, Andrew Crommelin of the Royal

Greenwich Observatory, would observe the eclipse from the city of Sobral in northeast Brazil.

The two parties left together, sailing from Liverpool on March 8 and arriving at the island of Madeira six days later. From there, the parties separated. Crommelin immediately found a ship to take him, his assistants, and his equipment to Brazil. Eddington was unsure whether he would find passage, this being just a few months after the signing of the armistice that ended the First World War, and the scheduling of ships still considered secret. Finally, after a wait of four weeks, a ship did appear, the S.S. *Portugal*. And Eddington and his assistants and equipment arrived at Principe on April 23, five weeks before the eclipse.

Eddington quickly discovered that Principe had terrible weather. Heavy rain was the norm. And whenever the rain abated, there were mosquitoes to contend with. And if one managed to cope with the rain and the mosquitoes, monkeys were a menace. At anytime, one or more would steal into camp and try to pilfer some small piece of equipment.

The morning of the eclipse, the weather worsened. It was the most tremendous rainstorm yet. The rain stopped at noon, just two hours before the eclipse, but the sky remained overcast.

When totality began, the dark disk of the Moon surrounded by the corona was visible through the clouds. At this point, as Eddington described it, there was nothing else to do "but to carry out the arranged program and hope for the best."

"I did not see the eclipse, being too busy changing plates, except for one place to make sure it had begun and another half-way through to see how much cloud there was." They took sixteen photographs, ranging from two to twenty seconds. Twelve plates would not show stars. Two of the plates showed only two stars. The remaining two plates each showed five stars. After the eclipse ended and the plates were developed, Eddington made a preliminary measurement of star positions. There seemed to be a deflection, but how much was yet to be determined. Eddington sent a telegraph to Dyson in London. "Through cloud. Hopeful."

Crommelin at Sobral did much better. He set up his equipment next to the grandstand of a racecourse of the Jockey Club at Sobral; tickets were sold to local people who wanted to look through the telescopes at night. They were given the use of Brazil's first automobile to ferry themselves around. A local meatpacker supplied them with ice that was used to control the temperature of chemicals that were used to develop the photographic plates. And they had excellent weather. Crommelin and his assistants managed to take seven photographs, some showing as many as seven distinct stars. Crommelin made no preliminary measurements of the plates. But he did send a telegraph to Dyson: "Eclipse splendid."

Though back in England by July, months would pass as careful measurements were made of star positions on the eclipse plates and on comparison plates, that is, on plates exposed to the same region of the sky without the Sun present. On November 6, 1919, at a joint meeting in London of the Royal Society and the Royal Astronomical Society, the results were announced.

Dyson, Astronomer Royal, spoke first. "After a careful study of the plates, I am prepared to say that they confirm Einstein's prediction." He continued. "The results of the expeditions to Sobral and Principe leave little doubt that a deflection of light takes place in the neighborhood of the sun and that it is of the amount demanded by Einstein's generalized theory of relativity."

Eddington followed. He noted that the deflection of starlight by the Sun's gravitational influence was the *second* confirmation of Einstein's theory of general relativity, the first one being Einstein's explanation of the precession of Mercury's perihelion. He also showed the results of the measurements. The photographic plates taken at Principe indicated a deflection at the Sun's limb of 1.98 arc-seconds, while those taken at Sobral showed 1.61 arc-seconds. The average was 1.79 arc-seconds, close to Einstein's prediction of 1.74 arc-seconds.

The next day, November 7, *The Times* of London carried on its front page the headline: REVOLUTION IN SCIENCE/NEW THEORY OF THE UNIVERSE/NEWTONIAN IDEAS OVERTHROWN. In the subsequent article, the newspaper introduced its readers to "one of the most momentous, if not the most momentous, pronouncements of human thought." The writer of the article also noted "no one had yet succeeded in stating in clear language what the theory of Einstein really was."

The public's intense interest in Einstein and his work and the mysterious aura that continues to surround him began on that day.

————•————

Astronomers were now eager to confirm Eddington's work. Seven eclipse expeditions were sent to Australia in 1922, and three succeeded in photographing star fields near the Sun and showed that indeed starlight was deflected by the amount predicted by Einstein. Similar results were reported for total solar eclipses in 1929, 1936, 1947, 1952, and 1973.

The most precise confirmation of Einstein's theory, however, has not come from observing the deflection of starlight, but from observing the deflection of radio waves (which, like visible light, is a form of electromagnetic radiation) by the Sun's gravity. Such observations do not require eclipses. Instead, radio measurements can be done whenever the Sun passes in front of a distant cosmic radio source, such as a quasar. Such measurements have been done many times and show that the amount of deflection between predicted and measured values agrees to much less than one percent, a strong confirmation of general relativity. From this, new fields of astronomy have spawned that use the deflection of both visible light and radio waves to measure the mass of distant galaxies, and from that to determine the amount of dark matter in the universe.

A discussion about the importance of the 1919 solar eclipse usually ends here. But the impact extends far beyond astronomy. To me, this was the most important eclipse yet observed because it yielded answers to two fundamental questions that have been asked since the beginning of history: What is knowledge? And what is truth?

————•————

Karl Popper was seventeen years old and a student in Vienna when Eddington announced his eclipse observations and that he had confirmed Einstein's prediction of the deflection of starlight. Popper and his fellow students were thrilled at the result—as was much of the world—because it was unexpected and provided a basis for a new reality. A year later, in 1920, Popper was in an audience that heard Einstein give a lecture about his theory. "It was quite beyond my understanding," Popper would write. And, yet, he continued, it "had a lasting influence on my intellectual development."

Popper was intrigued by Einstein's theory because it deviated greatly from the fundamental outlook of Newton's theory that had been successful for centuries. Newton posited that a mysterious force—gravity—controlled both the movement of apples falling off trees and of the planets orbiting the Sun. It was a world based on laws that seemed obvious. Einstein proposed that space and time were a continuum and that the way apples and planets moved depended on the shape of that continuum. Most importantly for Popper, Einstein had made a specific prediction—the amount of starlight that would be deflected—for a phenomenon that no one even knew existed. As Einstein said in the lecture, his theory should be considered as wrong if it had failed in the prediction.

Thus, wrote Popper, "I arrived at the conclusion that the scientific attitude was the crucial attitude, which did not look for verification but for crucial tests; tests which could *refute* the theory test." It was from this that Popper decided that science and nonscience could be distinguished. That is, this was the means to decide what was knowledge and what was truth. It rested on whether a theory could be falsified.

Verification of a theory was easy, said Popper. Astrologers did it all the time, in that some of their forecasts were later proved to be true. But that did not validate astrology as a science. It was the ability to prove an idea *false* that was at the heart of Popper's work—and the inspiration he received from Einstein and the result of the 1919 eclipse. It was the ability to possibly prove a theory *false* that determined whether it should be considered scientific or nonscientific. To put it another way: A scientific theory could never be proven to be true; it could only be proven that it was not wrong.

This became the central idea to Popper's philosophy of science. It was what made him the philosopher whose work influences how scientific research is conducted today. In fact, Popper's idea about how to distinguish between science and nonscience—which he called "demarcation"—is so strongly embraced today that for someone to charge that a scientist is proposing a theory that cannot be falsified is just about as harsh a blow and insult as can be offered.

And it was this idea that was inspired by Eddington's confirmation of Einstein's bold prediction during the 1919 total solar eclipse.

CHAPTER THIRTEEN

The Glorious Corona

As in the soft and sweet eclipse,
When soul meets soul on lovers' lips.

—Percy Bysshe Shelley,
Prometheus Unbound, 1820

H e was the inventor of the decimal point. That is, he was the first to use a dot to separate the integer and fractional part of a number, and thereby to simplify long calculations greatly. That was in 1593, in a publication about how to use an astrolabe. A decade earlier, in 1582, he wrote a long treatise explaining how best to adjust the calendar so that Easter, which was gradually becoming a summertime fest, would return to being celebrated in the early spring. And long before either of these two accomplishments, this man, the Jesuit father Christopher Clavius, entirely through happenstance, had the extraordinary good fortune to be standing twice within the Moon's shadow.

His first view of a total solar eclipse was in 1560 when he was a university student in Portugal. His description of the event was a meager one. He made no mention of how the Sun appeared, but he does say that "stars appeared in the sky" and that it was "marvelous to behold" when birds suddenly stopped singing and returned to their roosts. That was all. Fortunately, his description of the second eclipse held something of scientific importance.

It was 1567 and Clavius was now an instructor in mathematics in Rome at a Jesuit college. On April 9 of that year, at midday as he recorded it, the sky darkened considerably. This time he does describe the Sun's appearance, saying that he saw that "a certain narrow circle was left that surrounded the whole of the Moon on all sides." At first one might think that this was an annular solar eclipse when the Moon was too far away from the Earth to obscure the Sun's disk completely, but modern calculations show that, as seen from Rome, this must have been a total solar eclipse. And that the "certain narrow circle" is the first, rather vague, description of the solar corona.

Another notable description of the corona, this one much more definite, came a century and a half later by Edmond Halley who observed a total solar eclipse from London. During the event, he noticed that "a luminous ring . . . pale whiteness or pearl in color" completely surrounded and was concentric with the Moon. Based on this single observation, Halley suggested that the corona must be the Moon's atmosphere illuminated from behind by the bright light of the Sun. This remained the conventional wisdom until the late-nineteenth century when photographs taken from several locations during the same solar eclipse showed that the corona did not follow along with the Moon nor could it be an illusion produced by the Earth's atmosphere. Instead, by then, it was clearly established that the corona must be part of the Sun, perhaps, a greatly extended atmosphere that surrounded the solar disk.

Improvements in photography—more sensitive film, larger format cameras—allowed better and better photographs to be taken. And these were studied to determine whether the brightness or the size of the corona changed between eclipses. And changes, indeed, were seen. In particular, photographs of the corona taken in 1869 and in 1878 by the same people who were using the same equipment during both eclipses—one may remember that both eclipses were seen from the United States—showed that the corona was noticeably brighter and the rays extended from a wider range of angles and to greater distances from around the Sun in 1869 than in 1878. Moreover, the claim was supported by visual observations. Samuel Langley, who watched and photographed the 1878 eclipse from Pikes Peak in Colorado, described the brightest part of the corona of that year as "a narrow ring—hardly more than a line" and said that it was not as bright as the one he had seen eight years earlier in 1869.

Such variations continued to be observed. Peaks in coronal brightness and in the number of streamers were seen in 1869, 1882, and 1896; minima in both occurred in 1878, 1889, and 1900. And so there seemed to be a cyclical pattern, one that repeated every decade or so. And there was another cyclical pattern already known in the Sun's behavior—a cycle of sunspots.

Sunspots are dark blemishes that appear on what appears to be a featureless surface of the Sun. The fact that the number of sunspots varied in a regular way was discovered by German astronomer Samuel Schwabe. For seventeen years, he scanned the Sun's disk every clear day, hoping to catch an unknown planet moving in front of the Sun. He of course found no such planet. He reported his results in 1844. And that caused others to take up the task of watching the Sun routinely. Soon it was clear that the number of sunspots cycled over a period of about eleven years. During minimum years there were days, sometimes weeks, when no spots were seen. During maximum years there was seldom a day when no sunspots were seen and often there were hundreds present.

It did not take long to correlate the sunspot cycle with the variation in coronal brightness and in number of streamers. A peak in sunspot numbers occurred in early 1870, just a few months after the 1869 eclipse, and a minimum in the number occurred in 1878 when Langley reported the corona was dim and had few streamers.

There was another correlation known with the number of sunspots. The Earth's magnetic field, which could be stable for many years, was suddenly erratic when the number of sunspots was high. That suggested that sunspots were related to the magnetic activity of the Sun.

And so it all came together. Sunspots were related to the Sun's magnetic field. More sunspots were correlated with a more active corona. Therefore, the corona was probably produced by the Sun's magnetic activity. But that raised a question among those who knew the history of solar eclipses. After 1715 there were many descriptions of the solar corona; however, before that date, there were few, the account by Clavius being one of them. Did this mean that the Sun had just recently become magnetically active?

Walter Maunder of the Royal Observatory in London was just one of those who looked into this question, doing his work in the 1890s. He found that though there had been repeated cycles of sunspots since 1715, there were practically no sunspots seen during the prior seventy years, that is, back to 1645. Then, before that year, there were several reliable reports of sunspots. And so the conclusion was clear. There had been a prolonged period of no sunspots—a period now known as the Maunder Minimum—that ended abruptly in 1715, after which, there have been numerous reports of sunspots and of the solar corona.

Added to this, the aurorae, the shimmering lights that occasionally appear in the night sky and that are more commonly known as the Northern Lights or the Southern Lights, are also connected to the number of sunspots, and, hence, to the Sun's activity. And reports of them are rare between 1645 and 1715, though common afterward. All of this supports the idea that, yes, in about 1715 the Sun's magnetic field increased in activity—and remained so.

But what about before 1600? That was before the invention of the telescope, which has been important in seeing sunspots and compiling their numbers. And that was before the Renaissance and the Enlightenment and subsequent eras when the writing of diaries and letters was more common and the recording of aurorae and solar eclipses would also have been more likely. Is there a way to determine how active the Sun was before that date? Yes, there is.

The amount of radiogenic carbon, that is, carbon-14, in the atmosphere is related to the level of solar activity. To get ahead of the story, the Sun is always sending out a stream of charged particles called the solar wind. As the solar wind flows past the Earth, it provides a partial shield against cosmic rays, which originate far outside the solar system. When solar activity is high, the solar wind is strong and provides better protection against cosmic rays. Some of the cosmic rays that reach the upper atmosphere collide with atoms of nitrogen and convert them to atoms of carbon-14. And so, when solar activity is high, the conversion rate of nitrogen atoms to atoms of carbon-14 is low, and so the amount of carbon-14 in the atmosphere is low.

A record of the annual variation of carbon-14 has been obtained from tree rings. Each ring represents the annual growth of a tree. By determining how much carbon-14 is in a tree ring, it is possible to determine how much carbon-14 was in the atmosphere during that particular year.* And when that is done, it confirms what Maunder discovered. There was a lesser amount of carbon-14 in the atmosphere, and, hence, more solar activity, since 1715 than before. It confirms the Maunder Minimum from 1645 to 1715 and shows an earlier minimum that lasted from about 1420 to 1530.

---

* Because carbon-14 is radioactive, an adjustment must be made of the radioactive decay of carbon-14. Because the decay rate is constant, this is a straightforward correction.

In practice, the amount of carbon-14 in tree rings can be used to determine the level of solar activity going back about a thousand years. And what about earlier periods? A different indicator is used to do that.

Each year a layer of snow is deposited in the Arctic and Antarctic regions. As the snow continues to accumulate, older layers are compressed by the weight of newer ones, so that a series of ice layers are formed. Drilling into the ice and removing the ice cores can retrieve a record of these layers. The cores are examined in detail, the annual layers within them identified, and within the layers microscopic bubbles are seen that contain air that was in the atmosphere at the time a particular layer was deposited.

Much has been learned about these cores and the samples of ancient air. The record goes back several thousand years. It shows changes in climate. It reveals the timing of catastrophic volcanic eruptions that blanketed the planet with a layer of volcanic ash. It also indicates the past activity of the Sun through another radioactive element, beryllium-10, which like carbon-14 is also produced by cosmic rays.

The record of beryllium-10 found in air samples in microscopic bubbles in ice cores also shows the Maunder Minimum and the earlier period of low solar activity that occurred from about 1420 to 1530. It also shows that solar activity was low for most of the last 3,000 years. And so it does not seem that many people who lived in the medieval or ancient ages had much of an opportunity, as we do, to see a bright solar corona with streamers, which might explain why there are few descriptions of the corona found in writings from these earlier ages. But there is much more told by this remarkable record.

There were three periods of high solar activity that occurred between the seventh and the third centuries B.C.E. Each one lasted about eighty years. And, because each probably consisted of multiple sunspots cycles, each one is known, informally, as a *grand maxima*.

Today we are living during such a grand maxima in solar activity. Though the current cycle of sunspots started in 1715, the last several cycles, especially those since 1940, have been unusually high in sunspot numbers. And that is the reason for the remarkable series of spectacular coronas that have been seen recently. Inevitably, that will end, and probably soon.

If the history of solar activity over the last few thousand years is an indication of what to expect for the future, then the current grand maxima has probably run its course, having already lasted nearly eighty years. In fact, there is an indication that it might end soon. The most recent cycle of sunspots, which peaked in 2014, had the lowest number of sunspots since 1906. The next minimum should come in 2019, followed by a much suppressed maximum in 2025. The question that all solar scientists—and eclipse watchers—are wondering is whether there will even be another cycle of sunspots. As mentioned, the current cycle started abruptly in 1715. And though we do not know the physical mechanism that accounts for changes in solar activity, one should not be surprised if the current grand maxima ends soon.

If true, it means the most spectacular solar coronas of our lifetime have already happened. The ones that our children and our grandchildren will see may be much subdued. In fact, some solar scientists have suggested the start of another Maunder Minimum—when only dim coronas will be seen—for at least the next hundred years.

––––––––

There are actually two coronas visible during a total solar eclipse, the "false" corona and the "true" corona.* The brighter one—the one that varies in brightness and structure and is the highlight of most photographs of total solar eclipses—is the true corona. When Langley was standing at the top of Pikes Peak in 1878 and remarked at how dim the corona looked compared to the one he saw in 1869, he was referring to the true corona. The two faint broad streamers

––––––––

* In scientific literature, they are known as the "F-corona" and the "K-corona." For this discussion, we can ignore these formal names.

that he noted extended to great distances from opposite sides of the Sun was the false corona.

The false corona is easier to understand, and so it will be discussed first. It consists of a disk of tiny solid particles—about the size of household dust—that is orbiting the Sun. These particles are part of a much larger mass of tiny particles that lie close to the plane of the Earth's orbit that give rise to a familiar nighttime feature known as the *zodiacal light.*

The zodiacal light is a diffuse white glow that can be seen hanging over the western horizon during the hour or so after the Sun sets or over the eastern horizon just before it rises. It is sunlight that has been scattered by the tiny particles. And so the false corona is not part of the Sun or its atmosphere and does not change in brightness or shape during a sunspot cycle. The origin of the tiny particles is still debated. No samples have yet been retrieved. The most promising explanation, based on the fact that the zodiacal light is coming from a region that extends far above and below the plane of the ecliptic, is that these particles probably originate from comets that have disintegrated—for some unknown reason— somewhere in the vicinity of Jupiter or beyond.

The nature of the true corona is much more complicated.

To start, the name *corona* was first applied to this feature in 1806 by Spanish astronomer José Joaquín de Ferrer who was visiting the United States and watched the total solar eclipse of that year from Kinderhook, south of Albany, in New York. In his report, he notes that the light coming from the corona was greater than that of a full moon. (The sunspot cycle was near a maximum in 1806, and so the true corona would have been brighter than usual.) He also noted that the blackness of the lunar disk was in striking contrast to what he described as "the luminous corona" that surrounded it. In 1842 Francis Baily, viewing a total solar eclipse from Italy, compared the "bright glory" of the corona that he saw to the halos that painters usually drew around the heads of saints. Many of those who observed the 1860 total solar eclipse from Spain, and were attempting to take photographs, mentioned how the detailed form of the corona was better seen with the unaided eye than through a telescope. All of these accounts, and others, were highly descriptive, but none of them pertained to what the corona might actually be. As already pointed out, the fact that the corona was part of the Sun was not conclusively shown until 1868. Until then, the only thing understood about the corona is that it must be extremely tenuous because stars were sometimes visible through it during eclipses.

The first real scientific breakthrough about the nature of the true corona came in 1869 when Charles Young and William Harkness, working independently, both identified a green line in the spectrum of the corona. Unable to associate this spectral line with any known chemical element, Young and Harkness followed the example of Janssen and Lockyer for helium and suggested that the green coronal line was due to a new chemical element, one that they named *coronium.*

Work at later eclipses revealed more coronal lines, and these, too, could not be identified with known chemical elements. Confidence in the existence of coronium gradually declined, especially by the 1910s when Mendeleev's periodic table of chemical elements was almost filled and there was little room for a new element. And that is where the matter stood until a star was seen to burst in 1933 in the constellation Ophiuchus.

Walter Adams of the Mount Wilson Observatory in California took a spectrum of the star. In the spectrum he was surprised to find five coronal lines, including the green line discovered by Young and Harkness. This meant that the chemical element responsible for those lines was not unique to the Sun. Next Walter Grotrian working in Potsdam, Germany, learned of Adams's results. His intuition told him that the physical environment around a bursting star was probably much more extreme than around a normal star. In particular, he suggested that the starlight collected by Adams had probably come from an environment much hotter and much more tenuous than that of the Sun. That meant the lines in the spectrum that Adams had collected were "forbidden lines," that is, spectral lines produced by atoms in extreme environments where most of the electrons that normally surround an atom have been stripped off. Grotrian communicated this idea to Swedish physicist Bengt Edlén who had been doing experimental work on the spectra of atoms at high temperatures. By 1941, after doing a long series of experiments, Edlén had the answer. The green coronal line discovered by Young and Harkness was not a new chemical element. Instead, it was highly ionized iron, that is, iron with half of its electrons removed. And then came the startling conclusion. From other spectral measurements, the temperature of the Sun's surface was already known to be about 6,000 degrees. In order to strip away half of the electrons from an atom of iron, the temperature of the corona must be about *2,000,000 degrees!*

This was a fantastic result. And it explained much about the corona. It explained why the true corona extended so far away from the Sun: The extremely high temperature meant individual atoms had more energy to overcome the Sun's gravitational force.* And it explained why the true corona has rays and loops: The high temperature meant electrons have been stripped away from all of the atoms and those free electrons could move along the magnetic lines of force, producing the structure. And it explained the origin of the solar wind: It was the heat of expansion of the million-degree corona that was driving away the free electrons and other particles.

But the high temperature also opened up a major question: Why is the true corona so hot? That continues to be one of the most important questions in solar physics today. And a myriad of ideas have been offered.

The Sun is a huge ball of gas—actually, a plasma—that is so hot that electrons have been stripped completely from atoms, so that only bare atomic nuclei exist. The Sun's deep interior is hotter still—caused by the fusion of hydrogen to helium, which powers

---

* If the temperature of the true corona was a few thousand degrees, then the free electrons in the true corona would not have enough energy to overcome the Sun's gravitational pull and rise high enough to be seen during a total solar eclipse.

the Sun—and it is that heat that causes the Sun's outer layer to churn, or convect. This convection of free electrons and bare atomic nuclei produces a complex pattern of magnetic fields. The convection can carry some of the magnetic fields above the photosphere where the lines of the field—think of the lines associated with a bar magnet—twist and tangle, and occasionally break. When one breaks, the energy stored in the magnetic field is released. This breaking of magnetic lines may be going on millions of times a second and could be the energy source that heats the corona.

Alternately, the convection also produces waves in the magnetic fields. And these waves, which carry energy, propagate up through the photosphere and into the corona and, in that way, heat the corona. Which mechanism, or perhaps another one, is true is still debated.

———•———

There is much to see and very little time to do anything during a total solar eclipse. It begins when the Moon's silhouette first touches the bright disk of the Sun, a moment that is known as *first contact.*

It takes more than an hour for the Moon to move across and completely obscure the Sun. During that hour or so, the part of the Sun that remains visible is the photosphere. It is the source of all visible light coming from the Sun. It is light from the photosphere that heats the Earth. It is also the part of the Sun that is usually considered as the "surface," though there is no solid part of the Sun. Even the photosphere is a very tenuous gas, more than ten thousand times less dense than the Earth's atmosphere at sea level. What appears to be a sharp edge to the Sun is an illusion caused by a rapid increase in the opacity of the gases of the photosphere, a change from almost clear to nearly opaque that occurs over a distance of a few miles. It should also be noted that it is within the photosphere that sunspots occur.

The amount of sunlight obviously decreases as the Moon continues to slide in front of the Sun, but because the human eye can adjust to the lesser light, the decrease is hard to notice until several minutes before totality. By then animals have started to respond. A gust of wind might be felt. Edmond Halley noticed such a wind in 1715 when a "Chill and Damp attended the Darkness" that caused "some sense of Horror" among the spectators. As the Sun disappears behind the Moon the ground begins to cool. That halts the rise of warm air from the ground, which causes a change in wind speed and a change in wind direction. There may also be a noticeable drop in air temperature that can be accompanied by the formation of fog. In contrast, there is no evidence that solar eclipses cause a change in cloud patterns, probably because the duration of an eclipse is too short to change the air temperature of the upper atmosphere.

As totality nears, things happen quickly, pulling one's attention in different directions. The Moon's shadow will soon pass over the ground, moving from west to east. The last rays of brilliant sunlight will soon shoot through valleys along the Moon's rim and form Baily's beads. The corona begins to appear, though at first very faint.

Then, at the moment of *second contact*, when brilliant sunlight is completely obscured, a bright red streak might be seen for a few seconds along that section of the Moon's rim that has just extinguished the Sun. This streak is the chromosphere, which lies above the photosphere and which was first clearly described by George Airy in 1842 during a solar eclipse that he saw from Italy. He was looking at the eclipsed Sun through a telescope and saw what to him looked to be a range of bright red mountains. Today these "mountains" are known as spicules, and in appearance—in photographs much more detailed than Airy could have seen—have been likened to choppy sea waves or to a prairie fire burning through windblown grass. Actually, they consist of millions of gas jets. To put the chromosphere in perspective, the gases of this layer are about 5,000 degrees hotter and a million times less dense than the gases of the photosphere.

Around the rim of the eclipsed Sun may be one or more solar prominences. These are also masses of hot gases, rising higher than the chromosphere, which take the form of arches or irregular clouds that are suspended at these high levels by intense magnetic fields, such as those associated with sunspots.

And then there is, as Francis Baily remarked, the "bright glory" of the corona. Unlike the Earth's magnetic field, the magnetic field of the Sun does not consist of a simple dipole. Instead, the Sun's magnetic field is comprised of many pairs of regions of opposite magnetic polarity that are produced by the complex convective pattern within the Sun. The streaks and rays of the corona are the patterns of free electrons in the Sun's atmosphere made visible because they scatter light that strikes them, like motes in a sunbeam. And because the concentrations of these free electrons are constrained by the contortions of the magnetic fields, the shape of the corona reflects that of the magnetic fields. For that reason, the corona has no permanent features. It is never still, but always in a state of continual change.

And, much too soon, the eclipse is over. The moment when sunlight returns, the Sun and the Moon are at *third contact*. Then, when the Moon finally moves completely off of the Sun, it is *fourth contact*.

The exact time when each of these four contacts will occur is what one finds listed when examining a modern eclipse prediction, a prediction that represents a culmination of intellectual curiosity and the accumulation of knowledge from around the world that took more than four millennia to produce. It is this understanding of the intricate workings of the universe that is one of the hallmarks of modern thought. It is the modern way that we link ourselves to the heavens.

Illinois, 2017

There will be a total eclipse this year; autumn some time.

—James Joyce, *Ulysses*, 1922

Nowhere else in the solar system is it possible to stand on a solid surface and see a total solar eclipse except on planet Earth. Mercury and Venus have no moons. The two moons of Mars, Phobos and Deimos, are too far away to cover the Sun completely, and so no total solar eclipses can be seen in a Martian sky. The myriad of moons that swing around Jupiter, Saturn, Uranus, and Neptune would seem to have the potential for at least one of the moons to just barely obscure the Sun as seen from the surface of another one, but so far no such precise arrangement has been identified. The moon Charon is so close to Pluto that it covers much of the sky and what, from that remote part of the solar system, is a very diminutive Sun, so no total solar eclipse can be seen from the frozen surface of Pluto.

It should also be pointed out that even from Earth, total solar eclipses will not continue forever. There will be a last one. That is because the Moon is receding away from the Earth, a consequence of ocean tides. And so the day will come, hundreds of millions of year in the future, when the apparent size of the Moon will always be too small to completely cover the Sun. Whatever types of beings then inhabit the planet will not see one of nature's most remarkable cosmic spectacles.

We, however, do have that pleasure. We do live in a time of eclipses and during a period of history when they can be predicted precisely. And so we can plan, years in advance, where to be and when to be there to see the exact alignment of the Sun and the Moon, or of the Moon and the Earth's shadow. In a sense, when we search out eclipses, we are aligning ourselves to be part of the cosmic coincidence.

On August 21, 2017, the eclipse track runs across twelve states of the United States, including five state capitals.* Just seven years later, on April 8, 2024, the track will cross sixteen states and two Canadian provinces. Inevitably, the two eclipse tracks must cross. And they do so in a town that, for the next several years, can rightly proclaim itself to be America's eclipse capital, Carbondale, Illinois, where in 2017 the duration of totality will be 2 minutes and 38 seconds (which is just 2 seconds short of the maximum for this eclipse) and where in 2024 the duration will

---

* These are Salem, Oregon; Lincoln, Nebraska; Jefferson City, Missouri; Nashville, Tennessee; and Columbia, South Carolina. The path of totality of the 2017 eclipse barely misses Boise, Idaho and Topeka, Kansas.

be an impressive 4 minutes and 8 seconds.* That is almost seven minutes of totality from the same easily accessible locality! But Carbondale will quickly lose its status as an eclipse capital when the next total solar eclipse passes over the United States on August 12, 2045, the track of that eclipse crossing over fifteen states, passing south of Carbondale and southern Illinois by about 250 miles. The new eclipse capital of the United States will then be the town of Ola in Arkansas. Its population today is about one thousand. On April 8, 2024, and again on August 12, 2045, tens of thousands of people will be standing in Ola, Arkansas.

And that brings us to the final point. Eclipses may be the product of an extraordinary cosmic coincidence and today may be precisely predictable, but what drives people to station themselves to witness such events—and why eclipses have a unique place in human history—is that they never fail to produce the feeling of awe. Whether it was Caesar or Constantine or Lord Mountbatten or Monet or almost anyone else who lived in the past, or who will exist in the future, it is assured that just as you and I continue to do, that person has looked up and seen an eclipsed Sun or an eclipsed Moon. And has marveled at the grandeur of the sight.

---

* For comparison, Samuel Mitchell, a former director of the Leander McCormick Observatory at the University of Virginia and one of the more persistent eclipse chasers of the twentieth century, traveled more than 50,000 miles over a period of thirty years to witness five total solar eclipses. The accumulated duration of totality that he saw was about twelve minutes.

Total Solar Eclipses Across the United States, 2000 - 2050

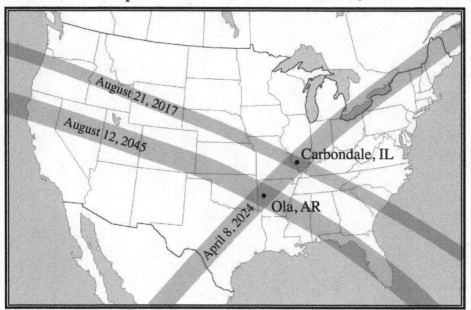

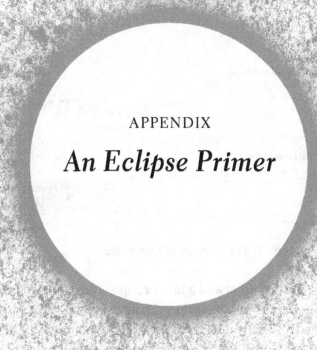

APPENDIX

An Eclipse Primer

Eclipses suns imply.

—Emily Dickinson, 1862

The basic fact about eclipses is this: A solar eclipse can occur only during a new moon and a lunar eclipse only during a full moon.

When the Moon passes between the Earth and the Sun so that the Moon's shadow falls on the Earth, there is a *solar* eclipse. When the Moon moves into the Earth's shadow, so that the Earth lies between the Sun and the Moon, there is a *lunar* eclipse.

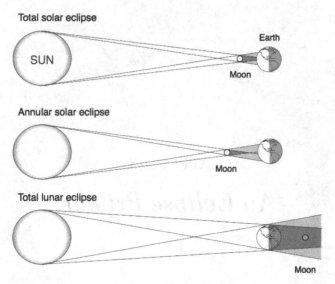

Total solar eclipse

Annular solar eclipse

Total lunar eclipse

The geometry of total eclipses.

There would be a solar and a lunar eclipse every month if the Sun and the Moon followed the same path across the sky. Instead, their paths differ. (A slightly more complicated explanation: the plane of the Moon's orbit around the Earth is inclined slightly to the plane of the Earth's orbit around the Sun. Where the two planes intersect is the line of lunar nodes.) And their paths differ in such a way that the Sun's path, known as the ecliptic, and the Moon's path cross each other at two points in the sky known as the lunar nodes. An eclipse is possible only when *both* the Sun and the Moon are near one of the nodes. If they are at the same node, then there is a solar eclipse. If they are at different nodes, then there is a lunar eclipse.

There are three types of solar eclipses, depending on how the Sun is obscured by the Moon. If, as seen by someone on the Earth, the Moon does not pass directly in front of the Sun, then there is a *partial* solar eclipse. And the Sun appears as a crescent. If the Moon does pass directly in front of the Sun, then there are two possibilities, depending mostly on the distance between the Earth and the Moon.

The distance between the Earth and the Moon varies because the Moon's orbit around the Earth is elliptical. When the Moon is in the far part of its orbit, it appears slightly smaller than the Sun, and so, if it passes directly in front of the Sun, it cannot cover it completely. A thin ring of sunlight remains. This is an *annular* solar eclipse. When the Moon is close to the Earth, it appears slightly larger than the Sun, and there can be a *total* solar eclipse.

There are also three types of lunar eclipses. Because the Sun is not a pinpoint in the sky but an extended object, the shadow produced by the Earth consists of two concentric circles. The inner circle, known as the *umbra*, is the dark part of the shadow where, if one was standing within the umbra, the Sun would be obscured completely and no direct sunlight would be seen. Outside the umbra is a broad ring—the *penumbra*—where, if one was standing within the penumbra, only part of the Sun would be obscured and some direct sunlight would be seen.

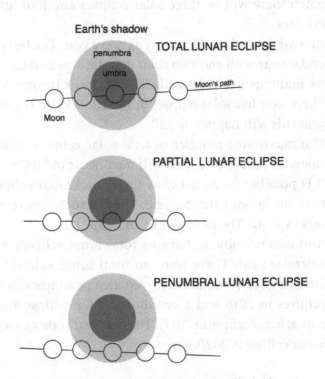

The Moon passing through the Earth's shadow
can cause three types of lunar eclipses.

If the Moon moves only through the *penumbral* part of the Earth's shadow, there is a penumbral lunar eclipse. If the Moon moves into the penumbra and only part of it is covered by the umbra, there is a *partial* lunar eclipse. And if the Moon moves through the penumbra and completely into the umbra of the shadow, then there is a *total* lunar eclipse.

Here are a few basic statistics about eclipses:

- At least *four* eclipses occur every twelve months, two solar and two lunar eclipses.

- As many as *seven* eclipses can occur in a calendar year. The most recent year when this happened was in 1982 when there were four solar eclipses and three lunar eclipses. The next year with seven eclipses will be 2038 when there will be three solar eclipses and four lunar eclipses.

- At least two solar eclipses occur every year. The last calendar year with *only* two solar eclipses was 2004.

- As many as five solar eclipses can occur in one year. There were five solar eclipses in 1805 and 1935. The next year this will happen is 2206.

- The maximum number of *total* solar eclipses in one calendar year is two. This will next occur in 2057.

- It is possible for as many as three total lunar eclipses to occur in one calendar year. The last such occurrence was in 1982. The next will be in 2485.

- And it is possible to have *no* total lunar eclipses in a calendar year. There were no total lunar eclipses in 2016 or 2017 (though there were two penumbral lunar eclipses in 2016 and a penumbral lunar eclipse and a partial lunar eclipse in 2017). The next year without a total lunar eclipse is 2020.

Acknowledgments

The mortal moon hath her eclipse endur'd

—William Shakespeare, *Sonnet 107*, c. 1605

P eople often watch eclipses in groups. I prefer to watch them in solitude. That is also how I chose to write this book.

That is not to say that there are no contemporaries who have influenced my ideas about eclipses. Foremost among them is Bruce Bills who, for many years, intrigued me during our almost daily discussions about how the Moon and the planets move and how they evolved. Richard Terrile tutored me on how to observe my first total solar eclipse. Through happenstance, Tom Peek ensured that I would have a clear and private view of the sky during my second one. And Wilfred Tanigawa, with whom I have had many fruitful discussions over the years, once stood his ground and was determined to see a total solar eclipse from the doorstep of his house, even though the weather was forecast to be cloudy. He did see the eclipse.

But I am indebted most to those who have been long departed. Among the ancients, there is Irassi-ilu, a servant of a Babylonian king, who was one of the first to provide written eclipse predictions and warn of the consequences. "When the Moon disappears on the thirtieth day," he once wrote in cuneiform, "there will be the clamor of the enemy in the land." The Roman historian and general Thucydides stands out for the three eclipses he recorded during the Peloponnesian War. Alfonso X of Castile had the foresight to have his mathematicians gather all available knowledge about the Moon's motion, and from that produce the first lunar tables, making it possible to predict eclipses routinely. A special mention must be made of Isaac Newton and the theory of lunar motion he published in 1707, and of Edmond Halley and the influence his map of the track of the 1715 total solar eclipse had on everyone who has ever wanted to see a solar eclipse. James Fenimore Cooper and Virginia Woolf left remarkable eyewitness accounts of viewing eclipses, as did Adalbert Stifter. I often think of Asaph Hall, who is known to history for the discovery of the Martian moons of Phobos and Deimos, who spent a month sailing across the Pacific Ocean from San Francisco to Kamchatka hoping to see a total solar eclipse, but the weather worked against him and he only saw the sky darken.

There are also three books that proved to be immensely helpful in the preparation of this book. *Total Eclipses of the Sun* by Mabel Loomis Todd gives an excellent account, in straightforward language, of the history and science of eclipses up to the end of the nineteenth century. *Total Eclipses of the Sun* by Jack Zirker updated Todd's book, also recalling to the reader the history and science of eclipses. And then there is *Eclipses of the Sun* by Samuel A. Mitchell. It has both a captivating text, including many firsthand accounts of eclipse watching, and a series of remarkable drawings, paintings, and photographs of eclipses.

There are two people who I am greatly indebted to for enabling me to write this book. Laura Wood, my agent, always brings the proper perspective to book writing. Jessica Case, my editor, has the forgiveness of a saint when I miss a deadline. It was through their efforts that this book was started and completed.

Sources

Was there an eclipse?

—Aristotle, fourth century B.C.E.

Hundreds of people have written about eclipses. The writings that follow are the subset that has influenced this book. Each one, in its own way, is a gem.

GENERAL REFERENCES

Cottam, Stella and Wayne Orchiston. *Eclipses, Transits, and Comets of the Nineteenth Century: How America's Perception of the Skies Changed.* New York: Springer, 2015.

Dreyer, J.L.E. *History of the Planetary Systems from Thales to Kepler.* Cambridge, England: Cambridge University Press, 1906.

Dyson, Frank and R.v.d.R. Woolley. *Eclipses of the Sun and Moon.* Oxford, England: Clarendon Press, 1937.

Kelley, David H. and Eugene F. Milone. *Exploring Ancient Skies: A Survey of Ancient and Cultural Astronomy.* Second edition. New York: Springer, 2011.

Lankford, John. *History of Astronomy: An Encyclopedia.* New York: Garland Publishing, 1997.

Levy, David H. *The Starlight Night: The Sky in the Writings of Shakespeare, Tennyson, and Hopkins.* New York: Springer, 2016.

SOURCES

Mitchell, Samuel Alfred. *Eclipses of the Sun*. New York: Columbia University Press, 1924.

Pasachoff, Jay M. "Solar eclipses as an astrophysical laboratory," *Nature*, 459, 789–795, 2009.

Schove, D. Justin. *Chronology of Eclipses and Comets AD 1–1000*. Dover, N.H.: Boydell Press, 1984.

Todd, Mabel Loomis. *Total Eclipses of the Sun*. Boston: Roberts Brothers, 1894.

Westfall, John and William Sheehan. *Celestial Shadows: Eclipses, Transits, and Occultations*. New York: Springer, 2015.

Zirker, Jack B. *Total Eclipses of the Sun*. New York: Van Nostrand Reinhold Company, 1984.

SELECTED REFERENCES

Prologue: New York, 1925

"Corona Splendor, Size Startled the Scientists," *Brooklyn Daily Eagle*, p.16, January 25, 1925.

"Eclipse Conditions Ideal, Flashes Giant Los Angeles," *Brooklyn Daily Eagle*, p.1, January 24, 1925.

"Flashes on Phone Trace 'Totality' Through the East," *Brooklyn Daily Eagle*, p.16, January 25, 1925.

Lamson, E. A. "The total solar eclipse of January 24, 1925," *Popular Astronomy*, 33, 523–526, 1925.

"Measurements of natural light during the solar eclipse of the sun, January 24, 1925," *Transactions of the Illuminating Engineering Society*, 20, No. 6, 565–628, July 1925.

"Most Perfect Eclipse, Declare Scientist on L.I.," *Brooklyn Daily Eagle*, p.16, January 25, 1925.

Pollock, Captain Edwin T., U.S.N. "Studying the eclipse from the *Los Angeles*," *McClure's Magazine*, 1, no. 2, new series, 9–22, June 1925.

"Scientific Observation of Eclipse on Dirigible *Los Angeles* Successful," *Standard Union* (Brooklyn, New York), p.1, January 24, 1925.

"Scientists on *Los Angeles* Praise First Dirigible Eclipse Flight," *New York Times*, p.1, January 25, 1925.

"Scientist View 'Celestial Glory' from Dirigible," *Brooklyn Daily Eagle*, p.4, January 24, 1925.

Chapter 1: The Heretic and the Pope

Anderson, Zoë. *The Ballet Lover's Companion*. New Haven, Conn.: Yale University Press, 2015.

Azzolini, Monica. "The political uses of astrology: predicting the illness and death of princes, kings and popes in the Italian Renaissance," *Studies in the History and Philosophy of Biological and Biomedical Sciences*, 41, 135–145, 2010.

SOURCES

Bonnerjea, Biren. "Eclipses among ancient and primitive peoples," *Scientific Monthly*, 40, no. 1, 63–39, 1935.

Campanella, Tommaso. *Selected Philosophical Poems of Tommaso Campanella: A Bilingual Edition*. Translated and annotated by Sherry Roush. Chicago: University of Chicago Press, 2011.

The Chronicle of John of Worcester: Volume III—The annals from 1067 to 1140 with The Gloucester Interpolations and The Continuation to 1141. Edited and translated by P. McGurk. Oxford, England: Clarendon Press, 1998.

Dooley, Brendan. *Morandi's Last Prophecy and the End of Renaissance Politics*. Princeton, N.J.: Princeton University Press. 2002.

Duncan, David Allen. "Campanella in Paris: Or, how to succeed in society and fail in the Republic of Letters," *Cahiers du Dix-septime: An Interdisciplinary, Journal*, 5, no.1, 95–110, 1001.

Ernst, Germana. *Tommaso Campanella: The Book and the Body of Nature*. Translated by David L. Marshall. New York: Springer, 2010.

Forshaw, Peter J. "Astrology, ritual and revolution in the works of Tommaso Campanella (1568–1639)," *The Uses of the Future in Early Modern Europe*. Edited by Andrea Brady and Emily Butterworth. London: Routledge, 181–197, 2009.

Gadbury, John. *Vox Solis: or, an Astrological Discourse of the Great Eclipse of the Sun, which Happened on June 22, 1666*. London: printed by James Cotterel, 1667.

Harrison, Mark. "From medical astrology to medical astronomy: Sol-lunar and planetary theories of disease in British medicine, c. 1700–1850," *British Journal for the History of Science*, 33, no. 1, 25–48, March 2000.

Hicks, Michael. *Anne Neville: Queen to Richard III*. Stroud, England: Tempus Publishing, 2006.

Horrox, Rosemary. *The Black Death*. Manchester, England: Manchester University Press, 1994.

Knobel, E. B. "On the astronomical observations recorded in the 'Nihongi,' the ancient chronicle of Japan," *Monthly Notices of the Royal Astronomical Society*, 66, 67–74, 1905.

Lepori, Gabriele. "Dark omens in the sky: do superstitious beliefs affect investment decisions?" *Copenhagen Business School Working Paper*, June 2009. (available at SSRN: http://dx.doi.org/10.2139/ssm.1428792)

Mayer, Thomas F. *Roman Inquisition on the Stage of Italy, c. 1590–1640*. Philadelphia: University of Pennsylvania Press, 2013.

Melvin, L. W. "Te Wahatoa of the Ngatihaua," *The Journal of the Polynesian Society*, 71, no. 4, 361–378, 1962.

Mitchell, Rose and Charlotte Johnson Frisbie. *Tall Woman: The Life Story of Rose Mitchell, a Navajo Woman, c. 1874–1977*. Albuquerque, N.M.: University of New Mexico Press, 2001.

Oestmann, Gunther, H. Darrel Rutkin, and Kocku von Stuckrad. *Horoscopes and Public Spheres: Essays on the History of Astrology*. Berlin: Walter de Gruyter, 2005.

SOURCES

Parascandola, John. *Sex, Sin, and Science: A History of Syphilis in America*. Westport, Conn.: Praeger, 2008.

Pope John Paul II. *On the Relationship between Faith and Reason. (Encyclical Letter Fides et Ratio of the Supreme Pontiff John Paul II to the Bishops of the Catholic Church on the Relationship between Faith and Reason)*. November 1998.

Rietbergen. Peter. *Power and Religion in Baroque Rome: Barberini Cultural Policies*. Leiden, Netherlands and Boston: Brill, 2066.

Sawyer, J.F.A. and F. R. Stephenson. "Literary and astronomical evidence for a total eclipse of the sun observed in ancient Ugarit on 3 May 1375 b.c.," *Bulletin of the School of Oriental and African Studies*, 33, no. 3, 467–489, 1970.

Scalzo, Joseph. "Campanella, Foucault, and Madness in Late-Sixteenth Century Italy," *Sixteenth Century Journal*, 21, no. 3, 359–372, 1990.

Snedegar, Keith. "Astronomical practices in Africa south of the Sahara," *Astronomy Across Cultures: The History of Non-Western Astronomy*. Edited by Helaine Selin. New York: Springer, 455–474, 2000.

Walker, D. P. *Spiritual and Demonic Magic: From Ficino to Campanella. Volume 22. Studies of the Warburg Institute*. London: The Warburg Institute, University of London, 1958.

Weech, William Nassau. *Urban VIII: Being the Lothian Prize Essay for 1903*. London: Archibald Constable & Co., 1905.

Yeomans, Donald K. "The origin of North American astronomy—seventeenth century," *Isis*, 68, no. 3, 414–425, September 1977.

Chapter 2: The Invisible Planets of Rahu and Ketu

Castleden, Rodney. *The Making of Stonehenge*. London: Routledge, 1994.

Friedlander, Michael W. "The Cahokia sun circles," *Wisconsin Archeologist*, 88, no. 1, 78–90, 2007.

Hamacher, Duane and Ray P. Norris. "'Bridging the Gap' through Australian cultural astronomy," *International Symposium on Archaeoastronomy & Astronomy in Culture*. Edited by Clive Ruggles, 282–290, 2011.

Hartner, Willy. "The pseudo planetary nodes of the moon's orbit in Hindu and Islamic iconographies," *Ars Islamica*, 5, no. 2, 112–154, 1938.

Hawkins, Gerald S. "Stonehenge decoded," *Nature*, 300, 306–308, 1963.

Hoyle, Fred. *On Stonehenge*. San Francisco: W.H. Freeman and Company, 1977.

Kaulins, Andis. "The Sky Disk of Nebra: evidence and interpretation." (available at: http://www.megaliths.net/nebraskydisk.pdf)

Kuehn, Sara. *The Dragon in Medieval East Christian and Islamic Art. Islamic History and Civilization, vol. 86*. Leiden, Netherlands and Boston: Brill, 2011.

Lawson, Andrew J. "The structural history of Stonehenge," *Proceedings of the British Academy*, 92, 15–37, 1997.

Malville, J. McK, R. Schild, F. Wendorf, and R. Brenmer. "Astronomy of Nabta Playa," *African Skies*, 11, 2–7, July 2007.

Markel, Stephen. "The imagery and iconographic development of the Indian planetary deities Rahu and Ketu," *South Asian Studies*, 6, 9–26, 1990.

SOURCES

Meller, Harald. "The Sky disc of Nebra," *The Oxford Handbook of the European Bronze Age*. Edited by Harry Tokens and Anthony Harding. Oxford, England: Oxford University Press, 2013.

Murray, William Breen. "Petroglyphic counts at Icamole, Nuevo Leon (Mexico)," *Current Anthropology*, 26, no. 2, 276–279, 1985.

———. "Numerical representation in North American rock art," *Native American Mathematics*. Edited by Michael P. Closs, 45–70, 1997.

———. "Boca de Potrerillos," *Handbook of Archaeoastronomy and Ethnoastronomy*. Edited by Clive Ruggles, 669–679, New York: Springer, 2015.

Pásztor, Emília. "Nebra disk," *Handbook of Archaeoastronomy and Ethnoastronomy*. Edited by Clive Ruggles, 1349–1356, New York: Springer, 2015.

Robbins, Lawrence H. "Astronomy and prehistory," *Astronomy Across Cultures: The History of Non-Western Astronomy*. Edited by Helaine Selin. New York: Springer, 31–52, 2000.

Ruggles, Clive "Current issues in archaeoastronomy," *Observatory*, 116, 278–285, 1996.

———. "Brainport Bay," *Ancient Astronomy: An Encyclopedia of Cosmologies and Myth*. Santa Barbara, Calif.: ABC-Clio, 48–50, 2005.

———. "Stonehenge and its landscape," *Handbook of Archaeoastronomy and Ethnoastronomy*. Edited by Clive Ruggles, 1224–1238, New York: Springer, 2015.

Smith, John. "Choir Gaur; The Grand Orrery of the ancient Druids, commonly called Stonehenge, on Salisbury Plain, astronomically explained and mathematically proved to be a Temple erected in the earliest Ages, for observing the Motions of the Heavenly Bodies," *Monthly Review*, 63, 353–357, 1770.

Smith, Walter. *The Muktesvara Temple in Bhubaneswar*. Delhi: Motilal Banarsidass Publishers, 1994.

Vahia, Mayank N. and B. V. Subbarayappa. "Eclipses in ancient India," *4th Symposium on History of Astronomy*. NOAJ, Japan, January 13–14, 2011. Edited by Mitsuru Soma and Kiyotaka Tanikawa. 2011.

Weiss, Peter L. "Reflections on refraction," *Science News*, 138, 236–237, October 13, 1990.

Chapter 3: Saros and the Substitute King

Aveni, Anthony. *Empires of Time: Calendars, Clocks, and Cultures*. New York: Basic Books, 1989.

———. *Stairways to the Stars: Skywatching in Three Great Ancient Cultures*. New York: John Wiley & Sons, 1997.

———. *Skywatchers: A Revised and Updated Version of Skywatchers of Ancient Mexico*. Austin, Texas: University of Texas Press, 2001.

Boardman, John and others. *The Cambridge Ancient History*. Second edition, vol. III, Part 2, *The Assyrian and Babylonian Empires and other States of the Near East, from the Eighth to the Sixth Centuries B.C.* Cambridge, England: Cambridge University Press, 1991.

SOURCES

Carman, Christián C. and James Evans. "On the epoch of the Antikythera mechanism and its eclipse predictor," *Archive for History of Exact Sciences*, 68, 693-774, 2014.

Cawthorne, Nigel. *Alexander the Great*. London: Haus Publishing Limited, 2004.

Edmunds, M. G. "An initial assessment of the accuracy of the gear trains in the Antikythera Mechanism," *Journal for the History of Astronomy*, 42, no. 3, 307-320, 2011.

Freeth, Tony. "Eclipse prediction on the ancient Greek astronomical calculating machine known as the Antikythera mechanism," *PLOS ONE*, 9, issue 7, 1-15, July 2014.

Harber, Hubert E. "Five Mayan eclipses in thirteen years," *Sky and Telescope*, 50, 72-74, 1969.

Jong, Matthijs J. de. "Chapter 5: Function of the Prophets," *Isaiah Among the Ancient Near Eastern Prophets*. Leiden, Netherlands and Boston: Brill, 2007.

Koetsier, Teun. "Phases in the unraveling of the secrets of the gear system of the Antikythera Mechanism," *International Symposium on History of Machines and Mechanisms*. Edited by H.-S. Yam and M. Ceccarelli, 269-294, 2009.

Krupp, E. C. *In Search of Ancient Astronomers*. New York: McGraw-Hill, 1978.

Lambert, W. G. "A part of the ritual for the substitute king," *Archiv for Orientforschung*, 18, 109-112, 1957/1958.

Meeus, Jean. "The frequency of total and annular solar eclipses for a given place," *Journal of the British Astronomical Association*, 92, no. 3, 124-126, 1982.

Nawotka, Krzyszlog. *Alexander the Great*. Cambridge, England: Cambridge Scholars Publishing, 2010.

Neugebauer, O. *The Exact Sciences in Antiquity*. Providence, R.I.: Brown University Press, 1957.

Parpola, Simo. *Letter from Assyrian Scholars to the Kings Esarhaddon and Ashurbanipal. Part II. Commentary and Appendices*. vol. 5, no. 2 in the series *Alter Orient und Altes Testament*. Verlag Button & Bercker, 1983

Sellers, Jane. *The Death of Gods in Ancient Egypt*. London: Penguin Books, 1992.

Timmer, David E. "Providence and perdition: Fray Diego de Landa justifies his inquisition against the Yucatecan Maya," *Church History*, 66, no. 3, 477-488, 1997.

Toth, Andrew. *Missionary Practices and Spanish Steel*. Bloomington: iUniverse, 2012.

Walton, John H. "The imagery of the substitute king ritual in Isaiah's Fourth Servant Song," *Journal of Biblical Literature*, 122, no. 4, 734-743, 2003.

Chapter 4: Measuring the World

Bleichmar, Daniela, Paula De Vos, Kristin Huffine, and Kebin Sheehan. *Science in the Spanish and Portuguese Empires, 1500-1800*. Stanford, Calif.: Stanford University Press, 2009.

Broughton, Peter. "Astronomy in seventeenth-century Canada," *Journal of the Royal Astronomical Society of Canada*, 75, 175-208, 1981.

SOURCES

Dugard, Martin. *The Last Voyage of Columbus: Being the Epic Tale of the Great Captain's Fourth Expedition, Including Accounts of Mutiny, Shipwreck, and Discovery.* New York: Little, Brown and Company, 2005.

Edwards, Clinton R. "Mapping by questionnaire: An early Spanish attempt to determine New World geographical positions," *Imago Mundi*, 23, 17–28, 1969.

Goodman, David C. *Power and Penury: Government, technology and science in Philip II's Spain.* Cambridge, England: Cambridge University Press, 2002.

Grimm, Florence M. *Astronomical Lore in Chaucer.* Lincoln: University of Nebraska, 1919.

Iqbal, Muzaffar. *The Making of Islamic Science.* Selangor, Malaysia: Islamic Book Trust, 2010.

James, Thomas. *The Strange and Dangerous Voyage of Captaine Thomas James.* London: John Legat, 1633.

Mackensen, Ruth Stellhorn. "Four great libraries of Medieval Baghdad," *The Library Quarterly: Information, Community, Policy*, 2, no. 3, 279–299, 1932.

Molander, Arne B. "Columbus's method of determining longitude: an analytical view," *Journal of Navigation*, 49, no. 3, 444–452, 1996.

Mundy, Barbara E. *The Mapping of New Spain: Indigenous Cartography and the Maps of the Relaciones Geograficas.* Chicago: University of Chicago Press, 1996.

Neal, Katherine. "Mathematics and Empire, Navigation and Exploration," *Isis*, 93, 435–453, 2002.

Neugebauer, O. *A History of Ancient Mathematical Astronomy.* New York: Springer, 1975.

Olson, Donald W. and Laurie E. Jasinski. "Chaucer and the Moon's speed," *Sky and Telescope*, 70, 376–377, 1989.

Pickering, Keith A. "Columbus's method of determining longitude," *Journal of Navigation*, 49, no. 1, 95–111, 1996.

Portuondo, Maria M. *Secret Science: Spanish Cosmography and the New World.* Chicago: University of Chicago Press, 2009.

———. "Lunar eclipses, longitude and the New World," *Journal of the History of Astronomy*, 40, 249–276, 2009.

Poulle, Emmanuel. "The Alfonsine Tables and Alfonso X of Castille," *Journal of the History of Astronomy*, 19, no. 2, 97–113, 1988.

Randles, W.G.L. *Portuguese and Spanish Attempts to Measure Longitude in the 16th Century.* Coimbra, 1984.

Rosen, Edward. "The Alfonsine Tables and Copernicus," *Manuscripta*, 20, 163–174, 1976.

Sen, S. N. "Al-Biruni on the determination of latitudes and longitudes in India," *Indian Journal of the History of Science*, 10, no. 2, 185–197, 1975.

Smoller, Laura. "The Alfonsine tables and the end of the world: Astrology and apocalyptic calculation in the later Middle Ages," *The Devil, Heresy & Witchcraft in the Middle Ages: Essays in Honor of Jeffrey B. Russell.* Edited by Alberto Ferreiro. Leiden, Netherlands: Brill, 211–239, 1998.

Zinner, Ernst. *Regiomontanus: His Life and Work.* Translated by Ezra Brown. Amsterdam: Elsevier, 1991.

Chapter 5: The Waste of Yin

Ben-Menahem, Ari. "Cross-dating of biblical events via singular astronomical and geophysical events over the ancient Near East," *Quarterly Journal of the Royal Astronomical Society,* 33, 175–190, 1992.

Brickerman, E. J. *Chronology of the Ancient World.* New York: Cornell University Press, 1980.

Chou, Hung-Hsiang. "Oracle bones," *Scientific American,* 240, issue 4, 134–149, 1979.

Dong, Linfu. *Cross Culture and Faith: The Life and Work of James Mellon Menzies.* Toronto: University of Toronto Press, 2005.

Elman, Benjamin A. *On Their Own Terms: Science in China, 1550–1900.* Cambridge, Mass.: Harvard University Press, 2005.

Grafton, Anthony. *Defenders of the Text: The Traditions of Scholarship in an Age of Science, 1450–1800.* Cambridge, Mass.: Harvard University Press, 1994.

Grafton, Anthony. "Some uses of eclipses in early modern chronology," *Journal of the History of Ideas,* 64, 213–229, 2003.

Ho Peng Yoke. "Astronomy in China," *Encyclopaedia of the History of Science, Technology, and Medicine in Non-Western Cultures.* Edited by Helaine Selin. New York: Springer, 108–111, 1997.

Hung-hsiang Chou. "Oracle bones," *Scientific American,* 240, no. 4, 134–149, 1979.

Jami, Catherine. *The Emperor's New Mathematics: Western Learning and Imperial Authority During the Kangxi Reign (1662–1722).* Oxford, England: Oxford University Press, 2012.

Li Chi. *Anyang.* Seattle: University of Washington Press, 1978.

Liu, Ciyuan and others. "Examination of early Chinese records of solar eclipses," *Journal of Astronomical History and Heritage,* 6, no. 1, 53–63, 2003.

Lü, Lingfeng. "Eclipses and the victory of European astronomy in China," *East Asian Science, Technology and Medicine,* 27, 127–145, 2007.

Rudolph, Richard C. "Lo Chen-yu visits the Wastes of Yin," *Nothing Concealed: Essays in Honor of Liu Yü-yün.* Edited by Frederic E. Wakeman. Taipei: Ch'eng wen ch'u pan she, 1970. (Republished in *China Heritage Quarterly,* no. 28, December 2011.)

Salvia, Stefano. "The battle of the astronomers. Johann Adam Schall von Bell and Ferdinand Verbiest at the court of the celestial emperors," *The Circulation of Science and Technology: Proceedings of the 4th International Conference of the ESHS, Barcelona, 18–20 November 2010.* Edited by Antoni Roca-Rosell. Barcelona: SCHCT-IEC, 959-963, 2012.

Steele, John M. "The use and abuse of astronomy in establishing absolute chronologies," *La Physique au Canada (Physics of Canada),* 59, no. 5, 243–248, 2003.

SOURCES

Stephenson, F. Richard. "Eclipses," *Encyclopaedia of the History of Science, Technology, and Medicine in Non-Western Cultures*. Edited by Helaine Selin. New York: Springer, 275–277, 1997.

Stephenson, F. Richard and Louay J. Fatoohi. "The eclipses recorded by Thucydides," *Historia: Zeitschrift for Alte Geschichte*, 50, no. 2, 245–253, 2001.

Sun Xiachun. "Crossing the boundaries between heaven and man: astronomy in ancient China," *Astronomy Across Cultures: The History of Non-Western Astronomy*. Edited by Helaine Selin. New York: Springer, 423–454, 2000.

Udias, Agustin. "Jesuit astronomers in Beijing, 1601–1805," *Quarterly Journal of the Royal Astronomical Society*, 35, 463–478, 1994.

———. *Jesuit Contribution to Science: A History*. New York: Springer, 2015.

Wilcox, Donald J. *The Measure of Times Past: Pre-Newtonian Chronologies and the Rhetoric of Relative Time*. Chicago: University of Chicago Press, 1987.

Yun Kuen Lee. "Building the chronology of early Chinese history," *Asian Perspectives*, 41, no. 1, 15–42, 2002.

Zhang, Qiong. *Making the New World Their Own: Chinese Encounters with Jesuit Science in the Age of Discovery*. Leiden, Netherlands and Boston: Brill, 2015.

Chapter 6: A Request to the Curious

Christianson, Gale E. *In the Presence of the Creator: Isaac Newton and His Times*. New York: Free Press, 1984.

Cook, Alan. *Edmond Halley: Charting the Heavens and the Seas*. Oxford, England: Claredon Press, 1998.

Forbes, Eric Gray. "The life and work of Tobias Mayer," *Quarterly Journal of the Royal Astronomical Society*, 8, 227–251, 1967.

Franklin, Benjamin. *Benjamin Franklin: Writings*. New York: Library of America, 440–442, 1987.

Gingerich, Owen. "Eighteenth-century eclipse paths," *Sky and Telescope*, 62, 324–327, 1981.

Grant, Robert. *History of Physical Astronomy from the Earliest Ages to the Middle of the Nineteenth Century*. London: Robert Baldwin, Paternoster Row, 1852.

Grier, David Alan. *When Computers Were Human*. Princeton, N.J.: Princeton University Press. 2005.

Halley, Edmond. "Observations of the late total eclipse of the sun on the 2nd of April," *Philosophical Transactions of the Royal Society*, 29, 245–62, 1715.

Leadbetter, Charles. *A Treatise of eclipses for 26 years: Commencing Anno 1715. Ending anno 1740*. London: J. and B. Sprint, 1717.

Manuel, Frank. *A Portrait of Isaac Newton*. Washington, D.C.: New Republic Books, 1968.

Pasachoff, Jay M. "Halley as an eclipse pioneer: his maps and observations of the total solar eclipses of 1715 and 1724," *Journal of Astronomical History and Heritage*, 2, 39–54, 1999.

Petit, Edison and Seth Nicholson. "Lunar radiation and temperatures," *Astrophysical Journal*, 71, 102–135, 1930.

Rolfe, Gertrude B. "The cat in law," *The North American Review*, 160, no. 459, 251–254, February 1895.

Rothschild, Robert. "Where did the 1780 eclipse go?" *Sky and Telescope*, 63, 558–560, 1982.

———. *Two Brides for Apollo: The Life of Samuel Williams (1743–1817)*. iUniverse, 2009.

Steele, John M. *Ancient Astronomical Observations and the Study of the Moon's Motion (1691–1757)*. New York: Springer, 2012.

Wepster, Steven. *Between Theory and Observations: Tobias Mayer's Explorations of Lunar Motion, 1751–1755*. New York: Springer, 2010.

Westfall, Richard S. *Never at Rest: A Biography of Isaac Newton*. Cambridge, England: Cambridge University Press, 1983.

Chapter 7: The Annulus at Inch Bonney

Baily, Francis. "Communications," *Monthly Notices of the Royal Astronomical Society of London*, 4, no. 2, 15–20, 1836.

———. "Some remarks on the total eclipse of the sun, on July 8th, 1842," *Monthly Notices of the Royal Astronomical Society*, 5, 208–214, 1842.

Cooper, James Fenimore. "The Eclipse," *Putnam's Monthly Review*, 352–359, 14, issue 21, 1869.

Dhir, S. P. and others. "Eclipse retinopathy," *British Journal of Ophthalmology*, 65, 42–25, 1981.

Dobson, Roger. "UK Hospitals assess eye damage after solar eclipse," *British Medical Journal*, 319, no. 7208, 469, 1999.

Herschel, John F. W. "Memoir of Francis Baily," *Monthly Notices of the Royal Astronomical Society*, 6, November 1844.

Holland, Jocelyn. "A natural history of disturbance: time and the solar eclipse," *Configurations*, 23, no. 2, 215–233, 2015.

Olson, Roberta J. M. and Jay M. Pasachoff. "Comets, meteors, and eclipses: Art and science in early Renaissance Italy," *Meteoritics & Planetary Science*, 37, 1563–1578, 2002.

———. "Blinded by the light: solar eclipses in art—science, symbolism, and spectacle." *ASP Conference Series*, 441, 205–215, 2011.

Osmond, A. H. and others. "Retinal burns after eclipse," *British Medical Journal*, 1, no. 5223, 424, 1961.

Smart, Alastair. "Taddeo Gaddi, Orcagna, and the eclipses of 1333 and 1339." *Studies in Late Medieval and Renaissance Painting in Honor of Millard Meiss. Volume I. Text*. Edited by Irving Lavin and John Plummer. New York: New York University Press, 403–414, 1977.

Vaquero José M. and M. Vazquez. *The Sun Recorded Through History*. New York: Springer, 2009.

SOURCES

Verma, L. and others. "Retinopathy after solar eclipse 1995," *National Medical Journal of India*, 9, no. 6, 266–267, 1996.

Woolf, Virginia. *A Writer's Diary: Being Extracts from the Diary of Virginia Woolf.* Edited by Leonard Woolf. New York: Harcourt, Brace and Company, 1953.

Chapter 8: A Simple Truth of Nature

Common, A. A. and A. Taylor. "Eclipse photography," *American Journal of Photography*, 11, no. 7, 203–209, 1890.

De La Rue, Warren. "On the total solar eclipse of July 18th, 1860, observed at Rivalbellosa, near Miranda de Ebro, in Spain," *Philosophical Transactions of the Royal Society*, 152, 333–416, 1862.

Launay, Francoise. *The Astronomer Jules Janssen: A Globetrotter of Celestial Physics.* New York: Springer, 2012.

Le Conte, David. "Two Guernseymen and Two Eclipses," *The Antiquarian Astronomer. Journal for the Society for the History of Astronomy*, issue 4, 55–68, January 2008.

———. "Warren De La Rue—Pioneer astronomical photographer," *Antiquarian Astronomer. Journal for the Society for the History of Astronomy*, issue 5, 14–35, 2011.

Lockyer, J. Norman. "Notice of an observation of the spectrum of a solar prominence," *Proceedings of the Royal Society of London*, 17, 91–92, 1868.

———. "Spectroscopic Observation of the sun," *Proceedings of the Royal Society of London*, 17, 131–132, 1868.

Meadows, A. J. *Science and Controversy: A Biography of Sir Norman Lockyer.* Cambridge, Mass.: M.I.T. Press, 1973.

Nath, Biman B. *The Story of Helium and the Birth of Astrophysics.* New York: Springer, 2013.

Pang, Alex Soojung-Kim. "The social event of the season: solar eclipse expeditions and Victorian culture," *Isis*, 84, no. 2, 252–277, 1993.

Rothermel, Holly. "Images of the sun: Warren De la Rue, George Biddell Airy and celestial photography," *British Journal for the History of Science*, 26, 137–169, 1993.

Schlesinger, Frank. "Some Aspects of Astronomical Photography of Precision," *Monthly Notices of the Royal Astronomical Society*, 57, 506–523, 1927.

Sears, Wheeler M. *Helium: The Disappearing Element.* New York: Springer, 2015.

Winichakul, Thongchai. *Siam Mapped: A History of the Geo-body of a Nation.* Honolulu: University of Hawaii Press, 1994.

Chapter 9: Eclipse Chasers

Barker, George F. "On the total solar eclipse of July 29th, 1878," *Proceedings of the American Philosophical Society*, 18, 103–114, 1878.

Bingham, Millicent Todd. *Ancestors' Brocades: The Literary Debut of Emily Dickinson.* New York: Harper & Brothers Publishers, 1945.

SOURCES

Cody, John. *After Great Pain: The Inner Life of Emily Dickinson*. Cambridge, Mass.: Harvard University Press, 1971.

Coffin, J.H.C. *Reports of the Observations of the Total Eclipse of the Sun, August 7, 1869*. Washington, D.C.: U.S. Department of the Navy, 1885.

Colbert, Elias. *The Solar Eclipse of July 29, 1878*. Chicago Astronomical Society: Chicago: Evening Journal Book and Job Printing House, 1878.

Eddy, John A. "The Great Eclipse of 1878," *Sky and Telescope*, 45, 340–346, 1973.

Gay, Peter. *The Bourgeois Experience: Victoria to Freud. Education of the Senses*. Oxford, England: Oxford University Press, 1984.

Gordon, Lyndall. *Lives Like Loaded Guns: Emily Dickinson and Her Family's Feuds*. New York: Viking, 2010.

Jones, Bessie Zaban. *Lighthouse of the Skies. The Smithsonian Astrophysical Observatory: Background and History 1846–1955*. Washington, D.C.: Smithsonian Institution. 1965.

Jones, Bessie Zaban and Lyle Gifford Boyd. *The Harvard College Observatory: The First four Directorships, 1839–1919*. Cambridge, Mass.: The Belknap Press of Harvard University Press, 1971.

Kendall, Phebe Mitchell. *Maria Mitchell: Life, Letters, and Journals*. Boston: Lee and Shepard Publishers, 1896.

"Lady Observers," *Daily Denver Tribune*, p.4, July 30, 1878.

Martin, Wendy. *All Things Dickinson: An Encyclopedia of Emily Dickinson's World*. (two vols.) Santa Barbara: Greenwood Printing, 2014.

Mitchell, Maria. "The total eclipse of 1869," *Friends' Intelligencer*, 26, no. 38, 603–605, 1869.

Reports on Observations of the Total Eclipse of the Sun, August 7, 1869. Appendix. Compiled by Commodore B. F. Sands. Washington, D.C.: Government Printing Office, 1869.

Reports on the Total Solar Eclipses of July 29, 1878, and January 11, 1880. Washington, D.C.: Government Printing Office, 1880.

Sheehan, William. "The Great American Eclipse of the 19th Century," *Sky and Telescope*, 132, 36–40, 2016.

Todd, David P. *American Eclipse Expedition to Japan, 1887. Preliminary Report (Unofficial) on the Total Solar Eclipse of 1887*. Amherst, Mass.: The Observatory, 1888.

———. "Automatic photography of the sun's corona," *Popular Astronomy*, 421, 309–317, 1933.

Todd, Mabel Loomis. *Corona and Coronet*. Boston: Houghton, Mifflin and Company, 1898.

———. "In the Moon's shadow," *Harper's Weekly*, 50, no. 2562, 120–123, 1906.

———. *Tripoli the Mysterious*. Boston: Small, Maynard and Company, 1912.

Woff, Cynthia Griffin. *Emily Dickinson*. New York: Alfred A. Knopf, 1986.

Wright, Helen. *Sweeper in the Sky: The Life of Maria Mitchell, First Woman Astronomer in America*. New York: MacMillan Company, 1949.

SOURCES

Chapter 10: Keys and Kettledrums

Adair, James. *The History of the American Indians*. London: Edward and Charles Dilly, 1775.

Anderson, Rani-Henrik. *The Lakota Ghost Dance of 1890*. Lincoln: University of Nebraska Press. 2008.

Barragan, Deborah I., Kelly E. Ormond, Michelle N. Strecker, and Jon Weil. "Concurrent use of cultural health practices and Western medicine during pregnancy: exploring the Mexican experience in the United States," *Journal of Genetic Counsel*, 20, 609–624, 2011.

Benedict, Ruth Fulton. "A brief sketch of Serrano culture," *American Anthropologist*, new series, 26, no. 3, 366–392, 1924.

Boas, Frank. *The Mythology of the Bella Coola Indians. Memoirs of the American Museum of Natural History. Part 2*. New York: G.P. Putnam's Sons, 1898.

Branch, Jane E. and Deborah A. Gust. "Effect of solar eclipse on the behavior of a captive group of chimpanzees (*Pan troglodytes*)," *American Journal of Primatology*, 11, issue 4, 367–373, 1986.

Bright, Thomas, Frank Ferrari, Douglas Martin, and Guy A. Franceschini. "Effects of a total solar eclipse on the vertical distribution of certain oceanic zooplankton," *Limnology and Oceanography*, 17, no. 2, 296–301, 1972.

Buff, Rachel. "Tecumseh and Tenskwatawa: Myth, historiography and popular memory," *Historical Reflections*, 21, no. 2, 277–299, 1995.

Dillard, Anne. "Total Eclipse," *Teaching a Stone to Talk: Expeditions and Encounters*. New York: Harper & Row, 1982.

Edmunds, R. David. *The Shawnee Prophet*. Lincoln: University of Nebraska Press, 1983.

Elliot, J. A. and G. H. Elliot. "Observations on bird singing during a solar eclipse," *Canadian Field Naturalist*, 88, 213–217, 1974.

Frazer, James George. *The Golden Bough: A Study in Magic and Religion*. twelve volumes. New York: MacMillan and Co., 1919.

Gray, Thomas R. *The Confessions of Nat Turner*. Baltimore: Lucas Deaver, 1831.

Greene, John C. "Some aspects of American astronomy 1750–1815," *Isis*, 45, no. 4, 339–358, 1954.

"Hindus use total eclipse for rituals," *Eugene Register-Guard*, page 10A, October 25, 1995.

Jortner, Adam. *The Gods of Prophetstown: The Battle of Tippecanoe and the Holy War for the American Frontier*. Oxford, England: Oxford University Press, 2012.

Lawrence, T.E. *Seven Pillars of Wisdom: A Triumph*. New York: Doubleday, 1926.

Lévi-Strauss, Claude. *The Raw and the Cooked: Introduction to a Science of Mythology. I*. New York: Harper & Row, Publishers, 1964.

Masur, Louis P. *1831 Year of the Eclipse*. New York: Hill and Wang, 2001.

McIlwraith, Thomas Forsyth. *The Bella Coola Indians, Volume 1*. Toronto: University of Toronto Press, 1948.

Metevelis, Peter. *Myth in History. Volume 2 of Mythological Essays*. San Jose, Calif.: Writers Club Press, 2002.

SOURCES

Mooney, James. *The Ghost-Dance Religion and the Sioux Outbreak of 1890.* Washington, D.C.: Government Printing Office, 1896.

Poole, DeWitt C. *Among the Sioux of Dakota: Eighteen months experience as an Indian Agent.* New York: D. Van Nostrand, 1881.

Russo, Kate. *Total Addiction: The Life of an Eclipse Chaser.* New York: Springer, 2012.

Sahagún, B. *Florentine Codex, Book 3—The Origins of the Gods.* Translated by A.J.O. Anderson and C. E. Dibble. *School of American Research Monographs, number 14, Part IV.* Salt Lake City: University of Utah Press, 1953.

Sanchez, Oscar, Jorge A. Vargas, and William Lopez-Forment. "Observations of bats during a total solar eclipse in Mexico," *Southwestern Naturalist,* 44, no. 1, 112–115, 1999.

Shylaja B. S. and Geetha Kaidala. "Stone inscriptions as records of celestial events," *Indian Journal of History of Science,* 47, no. 3, 533–538, 2012.

Svangren, M. L. "Observations of the solar eclipse, July 28, 1851," *Monthly Notices of the Royal Astronomical Society,* 12, 43–72, 1852.

Tylor, Edward B. *Primitive Culture: Researches into the Development of Mythology, Philosophy, Religion, Art, and Custom.* Two volumes. London: John Murray, 1871.

Wheeler, William Morton, Clinton V. MacCoy, Ludlow Griscom, Glover M. Allen, and Harold J. Coolidge Jr. "Observations on the behavior of animals during the total solar eclipse of August 31, 1932," *Proceedings of the American Academy of Arts and Sciences,* 70, no. 2, 33–70, 1935.

Chapter 11: The Crucifixion and the Concorde

Espenak, Fred and Jean Meeus. *Five Millennium Catalog of Solar Eclipses: -1999 to +3000 (2000 BCE to 3000 CE),* NASA/TP-2008-214170, Greenbelt, Maryland: National Aeronautics and Space Administration, Goddard Space Flight Center, 2009.

Harkness, William. "Total solar eclipse, Aug. 19, 1887," *Sidereal Messenger,* 7, no. 1, 1–9, 1888.

Humphreys, Colin J. and W. G. Waddington. "Dating the Crucifixion," *Nature,* 306, 743–746, 1983.

Léna, Pierre. *Racing the Moon's Shadow with Concorde 001.* New York: Springer, 2016.

Morrison, Leslie. "The length of the day: Richard Stephenson's contribution," *New Insights From Recent Studies in Historical Astronomy: Following in the Footsteps of F. Richard Stephenson.* Edited by Wayne Orchiston, David A. Green and Richard Strom. New York: Springer, 3–10, 2015.

Ruggles, Clive. "The Moon and the crucifixion," *Nature,* 345, 669–670, 1990.

Schaefer, Bradley E. "Lunar visibly and the crucifixion," *Quarterly Journal of the Royal Astronomical Society,* 31, 53–67, 1990.

Schlesinger, Frank and Dirk Brouwer. "Biographical memoir of Ernest William Brown 1866–1833," *National Academy of Sciences, Biographical Memoirs, Sixth Memoir,* 21, 243–273, 1939.

SOURCES

Steel, Duncan. *Marking Time: The Epic Quest to Invent the Perfect Calendar.* New York: John Wiley & Sons, 2000.

———. *Eclipse: The Celestial Phenomenon that Changed the Course of History.* Washington, D.C.: John Henry Press, 2001.

Steele, John M. *Ancient Astronomical Observations and the Study of the Moon's Motion (1691–1757).* New York: Springer, 2012.

Stephenson, F. Richard. "Historical eclipses and earth rotation: 700 BC–AD 1600," *Highlighting the History of Astronomy in the Asia-Pacific Region. Proceedings of the ICOA-6 Conference.* Edited by Wayne Orchiston, Tsuko Nakamura, and Richard Strom. New York: Springer, 3–20, 2011.

Stephenson, F. Richard. *Applications of Early Astronomical Records.* New York: Oxford University Press, 1978.

Stevens, Albert W. "Photographing the Eclipse of 1932 from the Air," *National Geographic,* 62, no. 5, 581–586, 1932.

Stewart, John Q., and James Stokley. "Observations of the eclipse of 1937 June 8 from near the noon point," *Publications of the Astronomical Society of the Pacific,* 49, no. 290, 186–189, 1937.

Wahr, John. "The Earth's rotation rate," *American Scientist,* 73, 41–46, 1985.

Wilson, Curtis. *The Hill-Brown Theory of the Moon's Motion: Its Coming-to-be and Short-lived Ascendancy (1877–1984).* New York: Springer, 2010.

Chapter 12: Einstein's Error

Albrecht, Sebastian. "The Lick Observatory-Crocker Expedition to Flint Island," *Journal of the Royal Astronomical Society of Canada,* 2, no. 3, 113–131, 1908.

Baum, Richard P. and William Sheehan. *In Search of Planet Vulcan: The Ghost in Newton's Clockwork Universe.* New York: Springer, 1997.

Crelinsten, Jeffrey. "William Wallace Campbell and the 'Einstein Problem': An observational astronomer confronts the theory of relativity," *Historical Studies in the Physical Sciences,* 14, no. 1, 1–91, 1983.

Eddington, Arthur S. *Space Time and Gravitation: An Outline of the General Relativity Theory.* Cambridge, England: Cambridge University Press, 1921.

Einstein, Albert. "Einfluss der Schwerkraft auf die Ausbreitung des Lichtes (On the influence of gravitation on the propagation of light)," *Annalen der Physik,* 34, series 4, 898–908, 1911. [English translation in Lorentz, H. A., A. Einstein, H. Minkowski, and H. Weyl. *The Principle of Relativity: A Collection of Original Memoirs on the Special and General Theory of Relativity.* Dover Publications, 97–108, 1923.]

———. *Relativity: The Special and General Theory.* New York: Henry Holt, 1920.

Fontenrose, Robert. "In search of Vulcan," *Journal for the History of Astronomy,* 4, 145–158, 1973.

Glass, Ian S. "Chapter 7. Arthur Eddington: Inside the stars," *Revolutionaries in the Cosmos: The Astro-Physicists.* Oxford, England: Oxford University Press, 198–234, 2006.

Kragh, Helge. "'The most philosophically important of all the sciences': Karl Popper and physical cosmology," *Perspectives on Science*, 21, no. 3, 325–357, 2013.

Lequeux, James. *Le Verrier—Magnificent and Detestable Astronomer*. New York: Springer, 2013.

Levenson, Thomas. *The Hunt for Vulcan: . . . And How Albert Einstein Destroyed a Planet, Discovered Relativity, and Deciphered the Universe*. New York: Random House, 2015.

Le Verrier, Urbain. "Le Verrier's report on the solar eclipse of July 18, 1860, at Tarazona in Spain," *American Journal of Science and Arts*, 30, second series, 309–312, 1860.

Malville, J. McKim. "The eclipse expeditions of the Lick Observatory and the dawn of astrophysics," *Mediterranean Archaeology and Archaeometry*, 14, no. 3, 283–292, 2014.

Moyer, Donald Franklin. "Revolution in science: the 1919 eclipse test of general relativity," *On the Path of Albert Einstein*. Edited by Berham Kursunoglu, 55–101, 1979.

Osterbrock, Donald E., John R. Gustafson, and W. J. Shiloh Unruh. *Eye on the Sky: Lick Observatory's First Century*. Berkeley: University of California Press, 1988.

Pearson, John C., Wayne Orchiston, and J. McKim Malville. "Some highlights of the Lick Observatory solar eclipse expeditions," *Highlighting the History of Astronomy in the Asia-Pacific Region. Proceedings of the ICOA-6 Conference*. Edited by Wayne Orchiston, Tsuko Nakamura, and Richard Strom, 243–338, New York: Springer, 2011.

Popper, Karl. *Conjectures and Refutations: The Growth of Scientific Knowledge*. New York: Basic Books, 1962.

———. *Unended Quest: An Intellectual Autobiography*. London: Open Court, 1982.

Rydin, Roger A. "The theory of Mercury's anomalous precession," *Proceedings of the NPA*, 8, 501–506, 2011.

Stanley, Matthew. "'An expedition to heal the wounds of war' The 1919 eclipse and Eddington as Quaker adventurer," *Isis*, 94, no. 1, 57–89, 2003.

Tice, John H. "The supposed planet Vulcan," *Scientific American*, 35, no. 25, 389, 1876.

Chapter 13: The Glorious Corona

Adams, Walter S. and Alfred H. Joy. "The spectrum of RS Ophiuchi (Nova Ophiuchi No. 3)," *Publications of the Astronomical Society of the Pacific*, 45, no. 267, 249–252, 1933.

Amari, Tahar, Jean-Francois Luciani, and Jean-Jacques Aly. "Small-scale dynamo magnetism as the driver for the heating of the solar atmosphere," *Nature*, 522, issue 7555, 188–191, 2015.

Barnard, L., A. M. Portas, S. L. Gray, and R. G. Harrison. "The National Eclipse Weather Experiment: an assessment of citizen scientist weather observations," *Philosophical Transactions of the Royal Society A*, 374, issue 2077, 2016.

SOURCES

Billings, Donald E. *A Guide to the Solar Corona*. New York: Academic Press, 1966.

Claridge, George C. "Coronium," *Journal of the Royal Astronomical Society of Canada*, 31, no. 8, 337–346, 1937.

Dwivedi, Bhola N. and Kenneth J. H. Phillips. "The paradox of the Sun's hot corona," *Scientific American*, 284, issue 6, 40–47, 2001.

Eddy, John A. "The Maunder Minimum," *Science*, 192, 1189–1202, 1976.

———. "The case of the missing sunspots," *Scientific American*, 236, issue 5, 80–88, 1977.

———. "The historical record of solar activity," *Proceedings of the Conference on the Ancient Sun*. Boulder, Colo., 1979. Edited by R. O. Pepin, J. A. Eddy, and R. B. Merrill. *Geochimica et Cosmochimica Acta*. Supplement. 13, 119–134, 1980.

Edlén, Bengt. "An attempt to identify the emission lines in the spectrum of the solar corona," *Arkiv för Mathematic, Astronomi och Fysik*, 28B, 1–4, 1941.

Ferrer, José Joaquín de. "Observations of the eclipse of the sun, June 16th, 1806, made at Kinderhook, in the State of New-York," *Transactions of the American Philosophical Society*, 6, 264–275, 1809.

Fisher, R. R. "Optical observations of the solar corona," *Space Science Reviews*, 33, 9–16, 1982.

Golub, Leon and Jay M. Pasachoff. *The Solar Corona*. Cambridge, England: Cambridge University Press, 1997.

Grotrian, Walter. "Zur Frage der Deutung der Linien im Spektrum der Sonnennkorona," *Naturwissenschaften*, 27, 214, 1939. (English translation in K. R. Lang and O. Gingerich. *A Source Book in Astronomy and Astrophysics 1900–1975*. Cambridge, Mass.: Harvard University Press, 1979.)

Hahn, Michael and Daniel Savin. "Observational quantification of the energy dissipated by Alfen waves in a polar coronal hole: evidence that waves drive the fast solar wind," *The Astrophysical Journal*, 276, no. 2, 10 pp., 2013.

Lang, Kenneth R. *The Cambridge Encyclopedia of the Sun*. Cambridge, England: Cambridge University Press, 2001.

———. *Essential Astrophysics*. New York: Springer, 2013.

Maunder, E. Walter. "A prolonged sunspot minimum," *Knowledge: An Illustrated Magazine of Science*, 17, no. 8, 173–176, 1894.

Nesvorny, David and others. "Cometary origin of the zodiacal cloud and carbonaceous micrometeorites: Implications for hot debris disks," *Astrophysical Journal*, 713, 816–836, 2010.

Phillips, Kenneth J. H. *Guide to the Sun*. Cambridge, England: Cambridge University Press, 1992.

Schwabe, Heinrich and Hofrath Schwabe. "Sonnen—Beobachtungen im Jahre 1843," *Astronomische nachrichten*, 21, 234–235, 1844.

Steinhilber, Friedhelm and many others. "9,400 years of cosmic radiation and solar activity from ice cores and tree rings," *Proceedings of the National Academy of Sciences of the United States*, 109, no. 16, 5967–5971, 2012.

SOURCES

Stephenson, F. Richard, J. E. Jones, and L. V. Morrison. "The solar eclipse observed by Clavius in A.D. 1567," *Astronomy and Astrophysics*, 322, 347–351, 1997.

Tassoul, Jean-Louis and Monique Tassoul. *A Concise History of Solar and Stellar Physics*. Princeton, N.J.: Princeton University Press, 2004.

Usoskin, Ilya G., S. K. Solanki and G. A. Kovaltsov. "Grand minima and maxima of solar activity: New observational constraints," *Astronomy & Astrophysics*, 471, 301–309, 2007.

———. "A history of solar activity over millennia," *Living Reviews in Solar Physics*, 10, 1–94, 2013.

Young, Charles A. "Theories regarding the Sun's corona," *North American Review*, 140, no. 339, 173–182, 1885.

Epilogue: Illinois, 2017

Espenak, Fred and Jay Anderson. "Get ready for America's coast-to-coast eclipse," *Sky and Telescope*, 131, no. 1, 22–28, 2016.

Gonzalez, Guillermo. "Wonderful eclipses," *News and Reviews in Astronomy & Geophysics*, 40, 3.18–3.20, 1999.

Appendix: An Eclipse Primer

Meeus, Jean. *Mathematical Astronomy Morsels*. Richmond, Va.: Willmann-Bell, Inc., 1997.

Index

INDEX

INDEX

INDEX

INDEX

INDEX

INDEX

INDEX